Becoming Human

Life and Mind: Philosophical Issues in Biology and Psychology
Kim Sterelny and Robert A. Wilson, Series Editors

Becoming Human: The Ontogenesis, Metaphysics, and Expression of Human Emotionality, Jennifer Greenwood, 2015

The Measure of Madness: Philosophy of Mind, Cognitive Neuroscience, and Delusional Thought, Philip Gerrans, 2014

Evolution in Four Dimensions: Genetic, Epigenetic, Behavioral, and Symbolic Variation in the History of Life, revised and updated edition, Eva Jablonka and Marion J. Lamb, 2014

Cooperation and Its Evolution, Volume 1: Agents and Environments, Kim Sterelny, Richard Joyce, Brett Calcott, and Ben Fraser, editors, 2012

Cooperation and Its Evolution, Volume 2: Agents and Mechanisms, Kim Sterelny, Richard Joyce, Brett Calcott, and Ben Fraser, editors, 2012

Ingenious Genes: How Gene Regulation Networks Evolve to Control Development,Roger Sansom, 2011

Yuck! The Nature and Moral Significance of Disgust, Daniel Kelly, 2011

Laws, Mind, and Free Will, Steven Horst, 2011

Perplexities of Consciousness, Eric Schwitzgebel, 2011

Humanity's End: Why We Should Reject Radical Enhancement, Nicholas Agar, 2010

Color Ontology and Color Science, Jonathan Cohen and Mohan Matthen, editors, 2010

The Extended Mind, Richard Menary, editor, 2010

The Native Mind and the Cultural Construction of Nature, Scott Atran and Douglas Medin, 2008

Describing Inner Experience? Proponent Meets Skeptic, Russell T. Hurlburt and Eric Schwitzgebel, 2007

Evolutionary Psychology as Maladapted Psychology, Robert C. Richardson, 2007

Becoming Human

The Ontogenesis, Metaphysics, and Expression of Human Emotionality

Jennifer Greenwood

The MIT Press
Cambridge, Massachusetts
London, England

This book was set in Stone Sans and Stone Serif by Toppan Best-set Premedia Limited. Printed and bound in the United States of America.

Library of Congress Cataloging-in-Publication Data is available.

ISBN: 978-0-262-02978-0

10 9 8 7 6 5 4 3 2 1

For Irene Greenwood
In loving memory

Contents

Preface

There are a number of interesting debates in contemporary emotion theory and a similar number in philosophy of mind, but two tend to stand out: the nature–nurture debate in emotion theory and the intracranialist–transcranialist debate in philosophy of mind. Theorists in emotion theory argue that emotions are either predominantly inborn, biological, or "natural" devices or predominantly learned, cultural, or "nurtured" devices. Intracranialist theorists in philosophy of mind argue that cognition takes place entirely in the head, and transcranialists argue that it can and frequently does take place in cognitive systems that extend into the natural, technological, and sociocultural world. This book has an important contribution to make to both debates. It demonstrates clearly that the nature–nurture debate is unfounded; biological and cultural resources are deeply functionally integrated throughout the development process. It also demonstrates clearly that human emotional and language development is a transcranialist achievement; human ontogenesis takes place only in extended cognitive systems that include environmental, technological, and sociocultural resources.

This book tells the story, the quite wonderful story, of how each one of us becomes a full human being. It tells the story of how human brains are constructed and how these brains acquire their contents through massive epigenetic scaffolding. The process of becoming fully human takes some twenty years to (almost) complete. This should be unsurprising.

Consider the human newborn. This newborn is the most dependent of mammalian newborns; she is minimally equipped with emotional, linguistic, and cognitive precursor preadaptations that function as assistance- or attention-soliciting devices to attendant caregivers. She is barely sentient, sensitive only to absolute stimulus thresholds that signal internal homeostatic conditions (e.g., water depletion) and external environmental conditions (e.g., loud noises). In addition, she is born into the most complex and challenging physical, technological, and social environments on earth. Yet, in some twenty years' time, this creature will become an entirely independent, autonomous agent, fully sapient, who emotes, thinks, and communicates in ways typical of, or unique to, her species and culture. What this implies is that the human newborn is a highly efficient learning machine.

Investigation into the "design specification" for this human learning machine demonstrates its remarkable elegance and economy. It includes a neonatal repertoire of minimal inborn coarse-grained constraints and capacities, a perfectly complementary repertoire of coarse-grained maternal constraints and capacities, and a range of neurochemicals that fuel the interaction of these two repertoires. The equipment of minimal coarse-grained capacities in both partners allows their progressive and flexible epigenetic fine-tuning. This flexibility allows the infant to adapt to the increasing complexity of her physical and social worlds as a result of increasing mobility and sociality. Limiting inborn resources to a minimum ensures that the infant can attend to only a minimum number of stimuli and exercise only a minimum number of capacities. This ensures that she focuses only on such stimuli and capacities to learn, respectively, how to recognize them and to practice their exercise. The maternal constraints repertoire ensures that the mother learns to respond promptly and sensitively to the infant's assistance- and attention-soliciting needs. The mother is, of course, equipped with the full repertoire of human capacities that the infant must learn during development, and the mother's constraints repertoire ensures that she has the motivation to help the infant learn them.

Endogenous opioids and prosocial neurochemicals are released during the mundane caring and nurturing activities that continue day in, day out, during the first year and beyond. What would inevitably be construed as tediously repetitive and sometimes even unappealing caring activities are rendered delightful through the release of these neurochemicals. They induce strong prosocial and bonding feelings in both partners to ensure that a close, linguistically mediated relationship develops between them. The mutual delight and developing relationship ensure that the interaction continues for the length of the developmental period. These neurochemicals also excite intraneural genetic products into neurogenesis and circuit maturation, and this process ensures that energy-expensive brain tissue is laid down only as needed to support the development of new capacities. The repetitious caring activities trigger neurogenesis and relationship development that enable emotional development, language development, and, by implication, the development of symbolic thought. These three capacities develop concurrently and through the same developmental mechanisms.

The chapters that follow provide considerable detail relating to the structures and processes underpinning human development, and chapter 1 includes a full introduction to the issues to be addressed. The book has been written for what I hope will be a multidisciplinary readership, and I have carefully defined all technical terms to facilitate this. Inevitably, however, some chapters will be more challenging than others, particularly for nonphilosophers; these are likely to be chapter 3, on metaphysics, and chapter 6, on biosemantics. Chapter 3 deals with the nature of the structures and processes that underpin emotion and cognition and also on their location. Chapter 6 focuses on how mental states, for example, thoughts, acquire their content or meaning, what philosophers term intentionality or "aboutness." Chapter 3 is informed by Wilson's (2004) work on metaphysics, and chapter 6 by Millikan's (1984) biosemantic theory. I have tried particularly to make the views of these extraordinary philosophers accessible to a multidisciplinary readership.

Researching the human emotional ontogenetic process has given me the intellectual ride of my life. Once I was into development, the process became utterly exhilarating (if punctuated occasionally by moments of desperation). As I stated at the beginning, human emotional ontogenesis, as well as ontogenesis more generally, "our story," is a truly wonderful story.

Acknowledgments

I am indebted to a number of people who have encouraged me and contributed to the development of my scholarship over the years: Jessie Parnell of University College Hospital, London, who recognized the importance of university education for nurses; Dr. Mike Golby, Exeter University, who supervised my MEd and first introduced me to philosophy; Professor Peter Tomlinson, University of Leeds, who supervised my first PhD (in education) and encouraged my interest in both philosophy and cognition; Professor Rob Wilson, University of Alberta, Canada, who was my adviser in my pre-PhD qualifying years and has continued to be a mentor since. I am grateful to all four for the quality of their mentorship. I am also extremely grateful to associate professor Deborah Brown, University of Queensland, who supervised the development of the thesis that informs this monograph.

I am also grateful to the (former) Australian Federation of University Women, Queensland, Inc., Fellowship Fund Branch for awarding me the 2010 Rosemary Venton Commemorative Fellowship. This fellowship allowed me to concentrate full-time for a year on my research and to turbocharge its development.

I wish also to acknowledge Sage Publications and the Taylor & Francis Group for their kind permission to include material previously published, respectively, in the following titles:

Greenwood, J. 2012. Wide externalism and the roles of biology and culture in human emotional development. *Emotion Review* 14 (4): 423–431.

Greenwood, J. 2013. Contingent transcranialism and deep functional cognitive integration: The case of human emotional ontogenesis. *Philosophical Psychology* 26 (3): 420–436.

Material from these two articles has been included mainly in chapters 3 and 4 of this book.

Finally, I wish to acknowledge the patience and professionalism of Cathy Nielson, who prepared the manuscript for publication, and Phil Laughlin (Senior Acquisitions Editor) and Judy Feldmann (Senior Editor), both of MIT Press, who guided me through the publication process.

My sincere thanks to all.

1 Introduction and Chapter Outlines

1.1 Introduction

All the world's a stage
And all the men and women merely players:
They have their exits and their entrances;
And one man in his time plays many parts,
His acts being seven ages. As, first the infant,
Mewling and puking in his nurse's arms.
And then the whining schoolboy, with his satchel
And shining morning face, creeping like snail
Unwillingly to school. And then the lover
Sighing like furnace, with a woeful ballad,
Made to his mistress' eyebrow.

—Shakespeare, *As You Like It*

This monograph tells the story of how a mewling, puking infant becomes an unwilling schoolboy and, eventually, a ballad-creating lover, but it is also a story that demonstrates, possibly surprisingly, just how important mewling is to development. It is a story that focuses on the development of human emotionality, but—because they are so closely connected, indeed, often develop concurrently—it is also a story of how the mewling infant becomes an articulate natural-language speaker (and, by implication, a symbolic thinker). Human infants and children learn to emote, to think, and to speak largely through the same developmental mechanisms and at the same time. It is really a story, therefore, that demonstrates how exquisitely the human newborn is minimally prepared biologically for a life of learning, a life of culture.

The epigenetic development implied in the last sentence should not be construed as sequential, that is, as first genetic and biological, and second environmental and cultural. Such a construal is typical of current orthodoxy in philosophy and psychology of emotion, and this is precisely one construal this book has been written to challenge. Contemporary orthodoxy in emotion theory construes emotions as falling into two distinct groups, one being largely innate, the basic emotions, and the other being largely socially constructed, the higher cognitive emotions. In addition, current orthodoxy construes emotions as operating primarily in psychological economies, that is, as operating primarily to benefit the subject-emoter and, as such, as individualistic. Individualism (or internalism) in philosophy of mind views the structures and processes that exist to determine and support emotional ontogenesis and experience (i.e., the underpinning metaphysics), and, indeed, mentality more generally, as being located entirely within the individual agent, in her brain and nervous system. In this book, I also argue that both of these construals are mistaken. I argue that basic emotions and, subsequently, higher cognitive emotions develop from inborn emotion precursors (affect expressions) and, further, operate primarily in social economies to enable human social life, first through interpersonal regulation and subsequently through intrapersonal regulation. In light of this, I also argue that the structures and processes that determine and support emotional ontogenesis and experience, and mentality more generally, extend beyond the brain and nervous system of the individual emoter-cognizer into her social and physical environment. The book therefore offers a novel theory of emotional ontogenesis, within the context of human ontogenesis more generally, and a metaphysics consistent with it.

I begin the story with the broadest of brushes to enhance its accessibility to what I hope will be a multidisciplinary readership. I become more fine grained later in this chapter and, more especially, in the chapters that follow.

The human neonate is the most immature of mammalian neonates, especially in terms of neurological and motor development. Her neurological development at birth is just sufficient to support an extremely limited repertoire of primitive emotional (affective), communicative, and cognitive capacities. She depends completely on others to provide for her basic needs, such as nutrition, warmth, and intimacy. To ensure that such needs are provided for, the neonate's primitive inborn capacities function as signals of relative physiological and psychological well-being. They signal to attendant caregivers that something of salience to the infant's well-being is being experienced, for example, that she is cold, hungry, uncomfortable, and so on. I term these primitive capacities *assistance-soliciting* (or *producer*) *devices*. In addition, and as an additional safeguard, her primary caregiver (typically her mother) is preadapted with a complementary repertoire of dispositions and skills, collectively termed intuitive parenting skills, that enable the caregiver to diagnose what is ailing or pleasing her infant and provide whatever assistance is required to remediate or prolong it. I term these dispositions and capacities *assistance-providing* (or *interpreter*) *devices*. Both of these repertoires are preadaptations, that is, the results of an evolutionary process of natural selection. Natural selection thus equips the human neonate with a range of assistance-soliciting devices, and her primary caregiver with a wonderfully complementary range of assistance-providing devices, to ensure the infant's survival and well-being. This is the proper function of such devices, the function they were selected for.

This barely sentient creature, however, is born into the most complex environment on earth, typified by apparently accelerating physical, social, and technological change. Despite the complexity of the environment, within sixteen to twenty years, the infant has morphed into a fully sapient, independent agent who emotes, communicates, and thinks in ways unique to, or typical of, her species and culture. She responds appropriately to the continuously changing exigencies of her environment. This results from the continuous interaction of the preadapted constraint repertoires of neonates and primary caregivers that bootstraps this development.

The constraints (assistance-soliciting devices) with which neonates are preadapted are minimal in number and extremely coarse-grained or general-purpose. They include primitive affect expressions (e.g., unfocused crying, motor unrest) that point to or express *something* of salience to the infant's well-being to attendant caregivers. As such, these are ostensive-expressive devices (ostension = to point to). As general-purpose capacities, they do not signal what the *something* is; caregivers have to discover this for themselves and correct it. Inborn constraints also include certain sensorimotor competencies, such as a preference for human(like) faces and voices. Such inbuilt preferences enable the infant to focus on a limited range of stimuli in her environment, thus saving her from possibly overwhelming stimulation. These preferences also ensure that attention focuses on preferred stimuli to allow practice in recognizing and responding to them. A further and critically important sensorimotor competence is motor mimicry.[1] Motor mimicry facilitates the recognizing and responding process; it allows the infant to copy and learn, for example, facial expressions, gestures, and words. It is a process that progressively enables the fine-tuning of the infant's emotional responses and, subsequently, through mimicry of words, her linguistic capabilities. Inborn capacities also include coarse-grained species-typical behavior patterns (e.g., prespeech "punctuated" suckling) that provide opportunities for caregivers to shape behaviors and for the infant to practice them.

Primitive affect expressions are triggered by the registering of absolute stimulus thresholds, both internal and external, by two types of receptors with which the neonate is equipped. The interoreceptors, which are widely located within the body, register significant changes in homeostatic conditions, for example, when glucose or temperature levels rise or fall significantly. The elicited affect expression, such as unfocused crying, triggers the parental response to deal with it. Proprioceptors are located in muscles, viscera, and skin; they register where body parts are in space. Proprioceptors in facial muscles, in particular, are crucial to the learning of emotional expression. Infants have to learn how a smile feels (as well as what causes smiles and what they are called).

The repertoire of dispositions and constraints (assistance-providing devices) with which the primary caregiver is preadapted is perfectly complementary. Intuitive parenting skills are triggered by a physiological stress reaction that parents feel when their infant is distressed. They are alerted to the infant and focus on her. Caregivers experience a strong desire to identify the cause of the distress and rectify it; that is, a search-identify-correct reaction is triggered by the infant's distress. This is a coarse-grained assistance-providing device. While exploring possible causes of the infant's distress, the mother engages in a form of affect mirroring that emphasizes the main features of the facial and bodily response of the infant. The infant responds by mimicking the expression (and learns how it feels, through the operation of her proprioceptors). The mother accompanies this affect mirroring with a continuous commentary of baby talk or "motherese." This is typically high-pitched and rhythmical and consists of simple two- or three-word utterances. The commentary focuses on what might be causing the infant's distress and what to do about it. The feedback elicited from the infant signals or fails to signal a correct diagnosis and treatment, which ensures that, in the future, the mother will try the same response or a different one. This serves to fine-tune the mother's assistance-providing devices. In short, the continuous cycle of maternal-neonatal interaction ensures that, simultaneously, both infant assistance-soliciting devices and maternal assistance-providing devices are progressively fine-tuned. It also ensures that language and, by implication, symbolic thought are progressively and simultaneously acquired.

These cycles of interaction continue in some form throughout the protracted human developmental period. They are fueled by the release of prosocial neuropeptides and endogenous opioids that induce intense feelings of intimacy and delight; these feelings, in turn, progressively develop the very close, linguistically mediated social relationship that provides the context for the development of progressively more complex forms of emotional, (cognitive), and linguistic behavior. The same neuropeptides trigger intraneural genetic products into neurogenesis or maturation and connection. The release of these neuropeptides is triggered by the auditory, visual, tactile, and olfactory sensory stimulation afforded through the mundane caring interactions of mother and baby. The process ensures that the requisite neural machinery is developed only as required to support developing new capacities.

What the foregoing overview implies is that philosophers might inhabit worlds of blooming, buzzing confusion, but human neonates manifestly do not. Human neonates are born into what I term *tight worlds*, characterized by constraints, context, and continuity (the three Cs); human emotional ontogenesis and human ontogenesis more generally result from the continuous interaction of these three factors. It is the continuous interaction of constraints that constructs the context in which human infants and children progressively (continuously) develop.

It is important to recognize that continuity also includes developmental mechanisms. The same mechanisms, that is, the interaction of the two mutually complementary repertoires, operate in the development of full human emotionality from emotion precursors (affect expressions) and in the development of segmented speech from a range of nonlinguistic and linguistic precursors. It is the very tightness of the constraint repertoires and the most protracted, highly dependent childhood on earth that enable the development of fully flexible thought, intrapersonal control of emotion, and full natural language. This is not to suggest that nonhuman animals do not “emote” in a manner appropriate to their lifeways; indeed, it is their supposed similarity to one group of human emotions, namely, basic emotions, that led to the construal of homology between animal and human emotions. It is to suggest, rather, that full *human* emotionality is language dependent.

It has been recognized since the 1960s that a close, linguistically mediated social relationship is the most significant factor in the development of distinctly human capacities; if denied this sort of relationship, human neonates and children will fail to develop into normal human adults. Scholars from a range of disciplines have contributed explanatory insights into this phenomenon, but few have shared their insights across disciplines. I think the single-discipline approach is understandable but regrettable; cross-disciplinary interaction lends breadth to research, as I hope this book bears witness. It is informed by insights harvested from philosophy and psychology of emotion, developmental psychology, metaphysics, biosemantics, and psycholinguistics that have usefully cross-fertilized. In this respect, my approach is consistent with that of two giants in philosophy. It is consistent with De Sousa (2014), who suggests that philosophy of emotion is vastly enriched by the relevant empirical literature. It is also consistent with Millikan (1984, 2004), who eschews conceptual analysis and philosophical argument concerning abstractions in favor of theory construction concerning exactly how people interact with their environments to create meaning and reference in thought. Research like mine sets out to construct just such a theory.

As a result of my interdisciplinary approach, I have come to construe emotion precursors and their affect expressions as being at one end of a continuum of increasingly complex ostensive-expressive devices; emotion precursors and discrete emotions are all ostensive-expressive devices of differing levels of specificity and sophistication. In addition, I construe the species-typical behavior patterns of human neonates that are present at birth or develop soon afterward also as being at one end of a continuum of ostensive-expressive devices, which develop through gestures and pragmatic foundation to end in full natural language. Relatedly, I construe emotion precursors and species-typical behavior patterns as referentially opaque expressions of homeostatic and physiologic status, the referential opacity decreasing progressively as caregiver behavior shapes the development of discrete emotions and language. Increasing referential clarity is acquired in both basic emotions and higher cognitive emotions with the assistance of language. In terms of language, referential clarity increases in the development of gesture, pragmatic foundation, and, finally, semantic language devices. Continuity is also reflected in the development of the neural substrata on which this continuum of ostensive-expressive devices partly supervenes. Each stage on the continuum represents a move away from fixed to highly flexible responses to salience (or relevance) detectors. The continuum begins in an extremely limited range of biological salience detectors (e.g., homeostatic absolute stimulus thresholds) and ends in a huge range of learned and considerably less obvious salience detectors (e.g., linguistic strings). All these responses to salience detectors are, of course, salience expressive.

Given these views on ostensive-expressive devices, another feature of my theorizing is my insistence that they are all intentional devices, even if, like emotion precursors and species-typical behavior patterns, only minimally so. They are intentional devices because they coevolved with the devices that use or consume them to the benefit of both. It was Dawkins (1986) who observed that sometimes traits are so perfectly complementary and integrated that "design" has to be implicated.

What the foregoing discussion also implies is that I construe all human communication to be essentially inferential; linguistic coding is clearly involved, but it enters inference processes as evidence of meaning just like nonlinguistic evidence. I think inference (albeit of a very simple kind) is involved in all mammalian communication.[2] Vervet alarms, for instance, communicate to conspecifics that a predator is approaching, but the conspecifics have to see where the communicating vervet is looking to discover from which direction the predator approaches. They also have to look at her facial expression to see how close the predator is (i.e., how alarmed she is). Human natural language relies on inferencing processes, too, but much more sophisticated and complex inference processes. In addition, and relatedly, I construe all communication, in terms of both production and comprehension, as irredeemably context sensitive. What all of this leads to is my conviction that human emotionality, language, and thought are overwhelmingly learned capacities. I argue that the human neonates' very few biological preadaptations equip them perfectly for development in a social environment replete with sensory-perceptual stimulation and symbols.

My research has also led me to a rather atypical view of epigenetic behavioral scaffolding. Epigenesis can suggest that behavioral development involves a process of transformation from imperfect, immature precursors into the more perfect, more coordinated, and finely tuned behavioral repertoires of adults. This view is generally attended by a construal of scaffolding as external adult support, for example, when an adult holds the baby upright to enable "walking." The support is always unidirectional, that is, from adult or adult-provided physical support structure to child and, as such, is always controlled by the adult. Of course, human adults do provide this type of scaffolding, as my walking example demonstrates. This is not the only type of scaffolding they provide, however, or the most important. The type of scaffolding that is crucial to human development involves the interactions of neonatal and maternal constraint repertoires, noted earlier, which provides the context for the development of progressively more advanced forms of behavior (and the neural substrata on which they supervene) in *both* partners, but to differing degrees. Both partners are jointly involved in the interaction and its control. For ease of reference, I term external adult scaffolding of the "walking" variety *Scaffolding 1*, and the coconstructing interactive variety *Scaffolding 2*.

Given that my account includes terms such as *preadaptation, biological equipment*, and others, it is worth emphasizing that it reflects a merely minimalist construal of nativism and utterly rejects genetic determinism. Nativists are inclined to view genes as specific instructional causes of maturation with environmental factors featuring merely as background or permissive causes. This is not the position I take. My position is that neonates possess what Perovic and Radenovic (2011) term *biological prerequisites* of maturated neurological centers that are present at birth or, more likely, emerge postnatally. These maturated neurological centers enable learning and are the result of genetic, epigenetic, and environmental instructive influences, as my account of emotional and language development illustrates. My position, therefore, is consistent with that of Perovic and Radenovic (2011), who characterize development as including two "segments," namely, a maturational segment followed by a social segment. In the maturational segment, biological processes interact with epigenetic and environmental processes to produce neural circuits or "endogenous biological machinery" (Perovic and Radenovic 2011, 399). In the social segment, which kicks in after the prerequisite neural circuits have been sufficiently maturated, learning is enabled also via biological, epigenetic, and environmental factors. The point is that learning cannot take place until these centers are developed. The distinction, therefore, does not depend on the relative primacy of genetic, epigenetic, or environmental influences; rather, it depends on the distinction between the processes and capacities that require learning and those that do not. This entire book, in particular chapters 4 and 5, is entirely consistent with this view:

> Everyday activities and routines in which a caregiver regulates child's eating, sleeping, levels of arousal, and the like are environmentally as well as biologically based processes necessary for the development of brain centres responsible for language learning or social cognition. Such social environmental causes are indispensable for the development of the brain structures after the child is born, as are the inner genetic and epigenetic (biological/environmental) processes in the prenatal period. Even so, the brain structures necessary for learning are not a result of the learning process which is virtually non-existent before the development of these centres is finalised. (Perovic and Radenovic 2011, 413)

The development of these views has led to an increasingly radical externalism in my theorizing. I was recruited to First Wave Extended Mind (EM), a progeny of mainstream functionalist information processing, by Clark and Chalmers's (1998) landmark paper "The Extended Mind," and like a number of First Wave enthusiasts, I have morphed into Third Wave via Second Wave. First Wave EM was based on a parity principle, that is, on functional isomorphism, and the Second Wave was based on the principle of complementarity, both parity and complementarity relating to biological and nonbiological elements in cognitive states and processes. Third Wave theorists construe First and Second Wave theories as presupposing that internal and external resources are fixed and noninterchangeable (Kirchhoff 2012). In addition, these theorists also suggest that Second Wave presupposes that sociocultural resources augment but do not significantly transform the brain's representational capacities during development. Third Wave theorists deny this idea.

Theorists who favor the complementarity of internal and external resources, which include sociocultural and natural environmental resources, construe them as deeply functionally integrated and as playing complementary roles in extended cognitive systems. Complementarity, therefore, is clearly a matter of degree. My own research focuses on the extremely deep and complex functional integration of neural and sociocultural resources that actually build and furnish brains, and to this extent my argument here is broadly consistent with Third Wave.

According to Clark, in a number of publications, mind "leaks" or "spills" into world, an idea that implies that extension is unidirectional, from mind to world. Clark (1997) does acknowledge world-to-mind extension, too, notably in his construal of continuous reciprocal causation, but his "leaking" and "spilling" metaphors are what apparently inform the standard view of EM; see chapter 3 in this book. My research demonstrates a reciprocal world-to-mind extension, and because of this I have deliberately replaced the term *EM* with leading intracranialists' term of abuse "contingent transcranialism" (CT) to signal my rejection of an implied mind-to-world unidirectional extension. I argue that there is a world-to-mind extension, and I use human emotional ontogenesis and language development as my examples.[3] I argue that the synchronous modulation (Scaffolding 2) of neonate and maternal activity that develops during emotional ontogenesis and language development and results in neurogenesis and neural connection is a world-to-mind transcranial achievement. In light of the degree of deep functional integration (DFI), which includes (contingent) transcranial processes, I replace EM with DFI with CT.

The research has taken five of my ten years' training in philosophy, during which time changes in construals regarding emotion, metaphysics, and language processing, in particular, have apparently accelerated. I briefly outline these various construals and show how they relate to my own theorizing in the chapters that follow. The sequence of these chapters is, of course, deliberate. I separate my chapter on metaphysical realization and function to present the monograph in what I hope are digestible bite-sized chunks. In addition, function and biosemantics appropriately fit between chapter 5, "Out of the Mouths of Babes and Sucklings," and chapter 7, "Loose Talk, Tight Worlds."

1.2 Chapter Outlines

1.2.1 Chapter 2: Theories of Emotion

In this chapter, to contextualize my own position that follows, I review traditional and contemporary theories of emotion to highlight their strengths and limitations. I begin by describing what are termed basic emotions (BEs) and higher cognitive emotions (HCEs) and highlighting their purported differences in terms of neurology, physiology, and behavioral expression. I include considerable detail here; the purported distinction between BEs and HCEs has been a dominant focus in emotion theorizing and has led, inter alia, to apparently mistaken conclusions concerning the nature of the emotions and their natural-kind status (although this purported distinction has recently been challenged).

Emotions are generally considered to fall into two basic and quite distinct groups, namely, affect programs or basic emotions and higher cognitive emotions; this is the "received view" in emotional science (Barrett 2011, 400). Basic emotions (BEs) are characterized as involving relatively stereotypical physiological, behavioral, and expressive patterns, which are underpinned by specialized neural circuits, and as genetically inherited as a result of a protracted evolutionary history. BEs are typically activated, from a very early age, by simple unconditioned stimuli and exhibit a degree of modularity such that neither the unconditioned stimuli that elicit them nor the responses they automatically generate can be moderated by higher cognitive functioning. As well as being activated by unconditioned stimuli, BEs can also be activated by conditioned stimuli through learning. In contrast, HCEs are considered to lack many of these features (Griffiths 1997). They are considered to be uniquely human and thus lack a shared evolutionary history with other animals. They are also considered to lack stereotypical physiological, behavioral, and expressive features and to vary culturally; any characteristics they possess are learned through and relative to the culture in which they are embedded. While both BEs and HCEs are considered capable of responding to higher cognitive inputs, HCEs are considered to lack simple unconditioned elicitors.

These views nicely encapsulate the contemporary biology–culture or nature–nurture debate; BEs are biological in that they are genetically inherited, and HCEs are socioculturally determined in that they are learned. BEs and HCEs have traditionally been viewed as so distinct that they could not form a natural kind (or one type be more basic than the other).

More recently, however, these views have begun to be challenged. Griffiths (2003) and Griffiths and Scarantino (2009) acknowledge that some HCEs may have simpler, basic forms and may profitably be discussed within a common theoretical framework. Rather more strenuously, Jason Clark (2010) argues that BEs and HCEs are not distinct categories but serial homologues, which means, although Clark fails to claim this explicitly, that they do form a natural kind, and BEs are more basic. My own research also challenges the distinction between BEs and HCEs, albeit from an ontogenetic perspective. I argue that both BEs and HCEs evolve from the emotion precursors with which neonates are biologically preadapted and through exactly the same developmental mechanisms.

I proceed to discuss, in just enough detail to be intelligible, feeling theories, cognitive theories (including dimensional appraisal theories), and social constructionist views. I identify (the admittedly well-rehearsed) objections to each of these views and conclude that although each has merit, none is adequate to explain the full complexity of human emotionality. In addition, and as a result of their almost exclusive focus on the nature of adult emotion, these theories misconstrue both the nature and function of the emotions. My review then turns to three more recent contributions, namely, Zinck and Newen's (2008) multifactorial developmental theory, J. Clark's (2010) views on emotional serial homology, and Colombetti's (2007, 2014) enactivist theory. I point out where my views are similar and dissimilar to these more recent contributions.

1.2.2 Chapter 3: Metaphysics and Mind

Here I focus on the intracranialist–transcranialist debate. I begin by outlining the differences between embodied, embedded, and extended cognition (Extended Mind), emphasizing that extended cognition carries distinct metaphysical baggage; it views the vehicles of cognition or mental states as extending beyond the skin and skull of the cognizing agent. Extended Mind theorists, as the name implies, claim that the vehicles of content can extend beyond the skin and skull of the cognizer, whereas individualists or contingent intracranialists deny this. After almost a decade of heated debate, Extended Mind theorists and contingent intracranialists remain completely unmoved by one another's arguments; indeed, their respective positions are so entrenched that scholars are beginning to talk of stalemate. My view is that Extended Mind theorists have failed to disarm the intracranialist challenge because they have failed to explicate precisely the nature of the deep functional integration on which their claims to vehicle or locational externalism depend and, in light of this, provide compelling examples of it.

Intracranialists accept that much of cognition is heavily scaffolded by extracranial technological, natural environmental, and sociocultural resources, that is, by Scaffolding 1. It is the metaphysics of externalism with which they take issue. Indeed, by misconstruing the nature of the external scaffolding of cognition, intracranialists claim that Extended Mind theorists make a number of errors:

(i) They fallaciously conflate causal coupling with constitution.
(ii) They fail to recognize the mark of the mental.
(iii) They are methodologically mistaken.
(iv) They view mind always as extending unidirectionally, that is, as mind extending into world.

Beginning in this chapter and continuing in all subsequent chapters, I argue that intracranialists are mistaken on all four counts. My analytic tool for this discussion is that of metaphysical realization. I briefly contextualize my discussion of metaphysical realization with a short history of the individualism–externalism debate and proceed to explicate R. A. Wilson's (2004) views on context-sensitive realization. This systematic treatment of metaphysical realization permits an analytic precision that might otherwise not be achieved. It provides a context-sensitive view of realization, which is the only one compatible with the theory of emotional ontogenesis I develop. I apply this analytic tool to two types of augmentations that transcranialists claim can extend cognitive systems, namely, technological and natural environmental systems. I conclude the chapter with Rupert's (2004) comparative analysis of the hypothesis of extended cognition versus the hypothesis of embedded cognition; this provides a useful summary of the main points of contention between intra- and transcranialists, as well as further opportunity to press the transcranialist case.

1.2.3 Chapter 4: Mirror, Mirror … Human Emotional Ontogenesis

This chapter focuses on human emotional ontogenesis. It draws heavily on empirical research from developmental psychology and the relevant cognitive neurosciences to address the limitations of traditional views of emotionality. It identifies the range of constraints in affect expressions and sensorimotor competencies and intuitive parenting skills, respectively, with which neonates and primary caregivers are preadapted and how they interact. It demonstrates how caregivers shape discrete emotions, in both BEs and HCEs, starting from referentially opaque affect expressions.

The depth of the complementarity, complexity, and integration of these in the neonate's emotional developmental trajectory become evident during the extremely detailed analysis. I include this level of detail to preclude any intracranialist denial of DFI with CT in this example. I argue that the repertoires of constraints are perfectly complementary and enable the development of the very close, linguistically mediated social relationship on which emotional ontogenesis so crucially depends. This relationship provides the context in which infants learn which stimuli cause which responses, their differential phenomenologies, and what they are called or termed in the embedding culture.

This analysis provides grounds for my claims that (i) emotions are ostensive-expressive devices that evolved to enable mammalian, including human mammalian, social life; and (ii) the same developmental mechanisms (and relationship) underpin the development of basic emotions from emotion precursors and subsequently to higher cognitive emotions. It demonstrates clearly that human neonates are biologically prepared for culture and that Scaffolding 2 is operational in human emotional ontogenesis. The analysis also subverts previous claims that the origins and development of BEs and HCEs are so different that emotions cannot form a natural kind.

Chapter 4 is meant to highlight the depth of the functional integration of intraneural and external sociocultural resources in emotional ontogenesis for which both neonate and caregiver are genetically preadapted. My analysis thus shows that there is no causal coupling–constitution fallacy operational in my theorizing about emotional ontogenesis.

1.2.4 Chapter 5: Out of the Mouths of Babes and Sucklings

This is a second chapter on emotional ontogenesis, and given that full human emotionality in both BEs and HCEs is language dependent, I also address the acquisition of language (and, by implication, the attainment of symbolic thought). As such, the discussion demonstrates, again, by including the same level of detail as in chapter 4, and for the same reason, the depth of complexity, complementarity, and integration, that is, DFI with CT in human ontogenesis.

My analysis focuses on species-typical behavior patterns (STBPs), which are innate or develop soon after birth, and how these provide the basis for the development of language. I argue that the linguistic environment into which human neonates are born and the complementary repertoires of neonatal STBPs and maternal intuitive parenting skills provide the necessary and sufficient conditions for the construction of a pragmatic foundation into which a semantic lexicon will subsequently be incorporated.

The analysis demonstrates the singular importance of "punctuated" suckling, a type of suckling that is unique to our species, which sets up the basis for joint attention and the development of different types of gestures. I argue that the importance of both turn taking and joint attention (which is also unique to our species) is underestimated in the relevant literatures; turn taking provides the time to imitate and practice new behaviors, and joint attention is hugely influential in speech production and comprehension.

It also demonstrates the importance of "motherese" in establishing the child's pragmatic framework. Exaggeratedly succinct maternal utterances, which emphasize clause boundaries and the prosodic changes that occur naturally at them, and maternal mirroring (and correction) of the infant's or toddler's "words," enable the construction of the pragmatic foundation of communication. The child naturally imitates mother's interactive timing, clause boundary recognition, and the rhythms, stresses, and intonations associated with commands, requests, statements, and questions. These are repetitiously practiced day in, day out, in mother-child interaction. I argue that before the child can comprehend and produce her first word, she can engage in "conversation" with adults. She can distinguish different linguistic devices in the utterances of adults and produce them herself by varying the pitch, volume, and rhythm of her own vocalizations.

This chapter once again demonstrates the depth of functional integration of internal and external resources in language development, for which both neonates and caregivers are genetically preadapted. It reinforces my rebuttal, in chapter 4, of the alleged intracranialist causal coupling–constitution fallacy.

1.2.5 Chapter 6: From Evolution to Emotionese

The analyses in this chapter support my arguments for the DFI with CT of emotional ontogenesis and language development, but they also do something more (or more explicitly than previous analyses): they demonstrate the complexity of the DFI that operates between intraneural and external resources. In addition, whereas chapters 4 and 5 rebutted the causal coupling–constitution fallacy claim, this chapter also rebuts the claim that transcranialists fail to recognize the mark of the cognitive.

In previous chapters, I noted that affect expressions and species-typical behavior patterns of human neonates are the results of brute physiological processes that are interpreted as—that is, function as—signs of physiological or homeostatic status by preadapted caregivers. That this is universally the case points to a selectionist history; affect expressions and species-typical behavior patterns (STBPs) are produced by devices that were selected for in ancestral environments (and continue to be selected for) because they conferred (and continue to confer) a fitness advantage on their bearers. Given this history, the chapter focuses on function and biosemantics (a naturalistic theory of how mental states acquire their contents or meaning, i.e., their intentionality).

The theory of function adopted is that of Millikan (1984) on proper function; I discuss direct proper function, and because they are so explanatorily powerful with respect to biological and cultural devices alike (and thus fit so well with my theory), I also discuss relational, derived, and adapted proper functions. I also use Millikan (2004) to analyze the relationship of natural, intentional, and linguistic signs and explain how one can emerge from another through a ratcheting process. Natural signs carry locally recurrent natural information; the information carried is recurrent because of the real causal connection between sign and signified. Intentional and linguistic signs carry locally recurrent information, too, but the only information represented intentionally is that which is useful to its consumers. Perception through language, according to this account, is as natural as perception through any other sensory-perceptual modality.

Finally, intracranialists allege that transcranialists fail to recognize the mark of the cognitive, which, they suggest, consists in nonderived representations, possessing intrinsic intentionality, which they further claim are found only in brains. I argue, in the light of this chapter's analyses, that the cognitive is marked by organismal attention to environmental saliences. In addition, I argue that representations are defined by the way they function not just in the head but in wider cognitive systems, which include representation-producing devices, objects in the environment, and representation-consuming devices. The representation-producing device produces representations that run isomorphically to the environmental conditions as defined by a semantic mapping rule, and the representation-consuming device uses this rule to produce behaviors that will be effective only if these environmental conditions obtain. Representations are defined by their proper functions, that is, the functions they were selected to have in their evolutionary past. Intentionality is grounded in external proper relations between representations and their representeds, the term proper being defined in terms of evolutionary history of either the species or the evolving individual (or both). There is nothing in the head, just as such, that displays intentionality.

1.2.6 Chapter 7: Loose Talk, Tight Worlds

In this last research-based chapter, it is entirely fitting that continuity, context, and constraints should feature as strongly as in previous chapters. Continuity is reflected in the move from natural signs (affect expressions and STBPs) to intentional signs (affect expressions, STBPs, gestures, and a range of inflectional and intonational pragmatic features) in previous chapters to linguistic signs (literal, loose, and metaphoric strings) in this chapter. Continuity is also reflected in the processing of linguistic devices in two ways. First, exactly the same processing mechanisms are operational in the processing of literal, loose, and metaphorical expression. Second, the processing involved is like all mammalian communication processing; that is, it is primarily inferential because it is context dependent. Context constrains both language production and interpretation in this account. What this implies is that my approach eschews a construal of human communication as primarily one of encoding–decoding. I argue that the code model of communication massively underdetermines communication interpretation, although human communication frequently includes encoded strings. These, however, enter the inferential process as evidence of meaning in exactly the same way as nonlinguistic evidence. I describe in some detail how a coding model underdetermines meaning, and show how an inference model avoids it.

I argue that language production and comprehension are always context dependent, and the tighter the world that interlocutors jointly inhabit, the looser their talk can afford to be. Indeed, as the discussion in chapter 5 demonstrates, in the very tight worlds that infants and caregivers jointly inhabit, where the information perceived is almost always locally recurring natural information, talk is hardly needed at all.

Continuity is also reflected in the relevance-theoretic approach of Sperber and Wilson (1995), which I adopt; the entire monograph focuses on the various forms that the expression of salience, that is, relevance, can take, so it is fitting that it concludes with this. A relevance-theoretic approach (obviously) construes verbal communication as governed by a principle of relevance and not the conventional construal of governance by maxim, rule, or convention of literal truthfulness. Truthfulness matters in communication, but only to the extent to which it is relevant to the needs of hearers and speakers.

I conclude the chapter by claiming that all verbal communication is widely realized and, to the extent that utterance activity and saccadic tracking are involved, radically widely realized.

1.2.7 Chapter 8: Once More, with Feeling

In the final chapter, I sum up the significant contributions of this research to theorizing in philosophy of emotion and human development more generally. I also suggest avenues for future research that arise from these contributions in both content areas and methodology. I suggest that, certainly with respect to human development, an intracranialist metaphysics should be abandoned. Such a metaphysics, as analyses in preceding chapters demonstrate, fuels if not actually generates the putative dichotomy of biology and culture, nature and nurture, which, in turn, underpins the mistaken construals of the BE-HCE distinction; it also fails to recognize sufficiently the continuity of human evolutionary development with that of other, nonhuman animals. I briefly summarize the results of previous chapters that claim that, inter alia, emotions are not primarily cognitive and somatic-feeling devices and do not function primarily in individual psychological economies. Emotions are essentially ostensive-expressive devices that evolved to regulate human social life, initially through interpersonal, and subsequently through intrapersonal, emotional regulation. In addition, analyses in previous chapters demonstrate that human emotions develop concurrently with, and through the same developmental mechanisms as, natural language (and, by implication, symbolic thought), and all three uniquely human capacities are radically widely metaphysically realized.

I suggest that avenues for future research include further investigation into the nature and operation of Scaffolding 1 and 2 and their role in education. I also suggest that moral development should be investigated; it is entirely plausible that children learn to think and behave morally as they learn to emote and to speak. I also suggest that further research is required into the extent of radically widely realized online interactivity.

I conclude by recommending, in light of the degree and complexity of the deep functional integration of neural, corporeal, and sociocultural resources in human ontogenesis, that a multidisciplinary, collaborative approach be preferentially adopted to investigate human development and behavior in naturalistic settings.

2 Theories of Emotion

2.1 Introduction

It is consensually agreed that emotions occur at the interface of our internal concerns and the outer world, especially our social world. As such, emotions are essentially what they are in virtue of their relational and functional properties, meaning that their properties can vary according to their relations and the context within which the agent finds herself. An agent can be terrified or mildly afraid of a whole range of objects and events, in a range of conditions that include her current physical and mental states, and a range of environmental, including social environmental, conditions. Emotions are evolved capacities (which may or may not be well designed for the modern world), many of which, scientists traditionally assumed, emerged from inbuilt neural circuits that are homologous in all mammals, others from repeated epigenetic interaction of those circuits with ecological and social environments, and yet others from our distinctly human capacity to semantically conceptualize issues of importance to us. At their most basic, emotions track affect valence and are associated with subcortical limbic structures in both human and nonhuman mammals.

Within both philosophy and psychology, emotions are considered to comprise or involve or essentially be the following:

- Thoughts (propositional attitudes, e.g., judgments, cognitive appraisals, desires) (Solomon 1977, 1993, 2003; Lyons 1980; Lazarus 1991; Arnold 1960)
- Modulation of mental processes (e.g., attention, memory) (Oatley 2004; Oatley and Johnson-Laird 1987)
- Categorization (L. Barrett 2005, 2006, 2009; Lindquist et al. 2012)

- Subjective feelings of pleasure or displeasure (Ledwig 2006)
- Subjective feelings of bodily changes: facial expression, orientation, musculature, visceral organs (Prinz 2004; De Lancey 2002; Griffiths 1997; le Doux 1998; Ekman 1984; Papanicolaou 1989; Zajonc 1980; Darwin 1872/1965; Lange 1885/1922; James 1884)
- Action tendencies (Frijda 1986)
- Interactive orientation devices (Zinck and Newen 2008)
- Paradigm scenario acquisition devices (De Sousa 1987)
- Imagination (Morton 2013)[1]

Although all theorists would agree that these components typify emotional experience, apart from the last one, they disagree on (at least) three issues. First, they disagree about which particular component is the emotion or is essential to emotional experience. Theories range from purely somatic-feeling (e.g., James 1884; Lange 1885/1922) to purely cognitivist (e.g., Nash 1989), with a range of less-radical views intervening, for example, somatic (Prinz 2004; Damasio 1994); and "hybrids" of various composites (e.g., Zinck and Newen's multifactorial theory [2008], Lyons's cognition-feeling hybrid [1980]). Second, theorists disagree about the extent to which emotions are genetically hardwired or culturally and socially constructed. Theorists who opine the former (e.g., Panksepp 1998, 2000, 2007; Ekman 1984, 2004) tend to do so because there is at least some evidence of universal emotion elicitors and expressions and because there is increasing evidence of specific neural substrata for at least some emotions (e.g., fear). Social constructionists, in contrast, deny the existence of universal emotions and therefore of genetic hardwiring. Emotions, to hard-nosed social constructionists (e.g., Harre 1986; Armon-Jones 1986; Averill 1980; Solomon 1977), are socioculturally determined. Third, theorists disagree on whether emotions are natural kinds and on their basicality or nonbasicality. Those who favor emotional natural kinds (e.g., Prinz 2004), or even just a "unified ontological class" (e.g., Zinck and Newen 2008), favor a blending or combination of basic emotions; those who reject the natural-kind status of emotions also tend to reject the blending or combination thesis.[2]

I address these three issues in this chapter. I review (of necessity, not exhaustively) the main theories of emotion, including those that relate to their nature and function, from both the philosophy and psychology of emotion. I show that even if subjective feelings and cognitions are necessary to individuate emotion (and there is evidence that, for some emotional capabilities anyway, they may not be), neither is sufficient. In addition, I show that despite their obvious dissimilarities, they share a number of less obvious similarities related to their foci. They all focus on fully developed adult emotions and on their essential nature, that is, their form. The debate focuses on the essentiality of cognitive judgments, propositional attitudes, perception, and bodily reactions and feelings in adult experience. This is despite consensus that emotions are evolved capacities, the functions of which increased hominid fitness. In addition, this focus on adult emotions persists despite agreement that the emotional repertoires of adult humans and human infants and children vary considerably in both number and complexity. As a consequence, both contemporary philosophies and psychologies miss the point of the emotions and misconstrue their essential nature. They do involve perception, cognition and judgment, and bodily reactions and feelings, to differing degrees and in different circumstances, but they are not essentially perceptive or cognitive devices. In the two chapters that follow, I show that the emotions are essentially ostensive-expressive, that is, communicative devices that evolved to regulate mammalian (including human mammalian) social life, first through interpersonal regulation, and subsequently through intrapersonal regulation. The focus on adult emotionality leads to a mistaken construal regarding the nature of basic emotions and higher cognitive emotions and, as a result, on their basicality or nonbasicality and natural-kind status. I conclude that analyses of emotional ontogenesis should reveal how an extremely limited range of patterns of bodily perturbation refers to, or is implicated in referring to, an extensive range of distinct perceptive-cognitive-affective-behavioral complexes. I begin, however, with some brief, rock-bottom preliminaries related to emotion.

2.2 Emotions: Some Rock-Bottom Preliminaries

The emotions are typically classified into (at least) two basic groups despite the recent challenge to this distinction; see chapter 1. These are affect programs (Griffiths 1997; Ekman 1973), also known as blue-ribbon emotions (Maclean 1990) and basic emotions (J. Clark 2010), and the higher cognitive emotions (Griffiths 1997), also known as higher sentiments (Maclean 1990). These two groups are a function of discernible stages in brain development, characterized by increasing neurobiological volume, complexity, and interconnectedness (Panksepp 1998, 2000, 2007).

2.2.1 Blue-Ribbon, Basic Emotions (BEs)

Basic emotions evolved from primitive reptilian brains, composed almost exclusively of corpus striatum, which still forms the innermost and oldest part of the brain. These reptilian brains controlled the essential life functions of land-based vertebrates, such as choice of home site, choice of mate, territory defense, flocking, and daily routines. Daily routines include waking, local foraging, defecating, returning home, and settling to sleep. Neural circuits arising in this ancient part of the brain control reflexive affects, a group of eruptive and transient responses that are closely tied to precipitating environmental conditions, such as startle, gustatory disgust, pain, and the various homeostatic distresses (e.g., hunger, thirst) and pleasures (e.g., good tastes).

Essential life functions and reflexive affects are governed by reactive representations that register states of the environment and trigger appropriate fixed responses. The connections between registration and response are simple and hardwired; the creature that possesses only this type of representation, therefore, is completely under environmental control (R. Wilson 2004).

Basic emotions, that is, those typically thought to include (but see Prinz [2004] for an interesting alternative view) fear, anger, sadness, joy, affection, and interest, can markedly outlast precipitating environmental (and, in humans, internal precipitating) conditions. They are governed by a set of neural circuits in intermediate areas of the brain (Panksepp 1998, 2000, 2007) and are conceptualized as sensorimotor command circuits.[3]

The cortices of mammals orchestrate the complex and coordinated behavioral, physiological, cognitive, and affective responses characteristic of fear, anger, and so on. In terms of complexity, they involve several expressive elements in, for example, changes in facial musculature, musculoskeletal functioning (e.g., flinching), orienting, expressive vocal changes, endocrine system changes, changes in autonomic nervous system (ANS) activity, subjective feelings, and a range of cognitive phenomena, such as directing attention. In terms of coordination, various elements co-occur in recognizable patterns or sequences, and in terms of automation, they unfold without need for conscious attention (Ekman 1984).

Blue-ribbon or basic emotions and their underlying neural circuits appear to be present in homologous fashion in all mammalian species, human and nonhuman, though their nature and expression could differ in fine detail (Griffiths 1997). The experience of most basic emotions (fear and interest excepted) presupposes sociality. Mammals take care of attached offspring (reptiles do not), they signal vocally and nonvocally to each other, and the young engage in rough-and-tumble play.

The local events that possess the properties of being dangerous, novel, and so on, will differ from one environment to the next; the young, however, inherit their environments along with their genes (Griffiths 1997, 61). Environmental influences, including parental guidance, inscribe properties of relevant stimuli. Learning is therefore involved in selecting which stimuli activate appropriate BEs, but this learning is, to greater or lesser extents (depending on the complexities of organismic nervous system and environment), biologically primed or constrained in that responses are attached much more easily to some stimuli than to others (Ohman 1986). Indeed, it has even been suggested that subpersonal organismic perceptual apparatuses are coupled to salient environmental particulars (O'Regan and Noe 2001), which, in effect, off-loads some representational burden onto the environment. Both of these are true of human neonates; I return to both points at the appropriate juncture.

The cortices of mammals, which make up 30 to 40 percent of their brain mass, enable more social learning (and therefore more flexible responses to environmental exigency) than their more primitive ancestors. Such relatively flexible responses are controlled by enactive representations that register the state of the environment but do not automatically generate a response. Registrations are combined before bodily responses are generated. Creatures that possess enactive representations perceive both external and internal stimuli and appropriate responses; their locus of control, therefore, is shared with the environment (R. Wilson 2004).

Neural Bases of Basic Emotions (BEs)

Blue-ribbon emotions are subserved by dual circuits involving a very rapid ("quick and dirty") "low road" and a much slower "high road" (Le Doux 1998, 2000). The low road consists of sensory inputs to the amygdala (Greek for "almond"), which appraises positive and dangerous stimuli, via thalamic projections both to expedite detection of environmental threats and to prime the amygdala to evaluate subsequent input arriving via the cortical circuit. The high road consists of circuits from sensory thalamus to sensory cortex and thence to amygdala.

In terms of an auditory stimulus "BANG":

(i) The low road

The sensory thalamus is activated by the intense auditory stimulus "BANG"; the thalamus activates the amygdala via neural projections, which in turn activate a variety of subcortical structures that elicit the ANS changes, motor commands, and facial expression and hormone changes characteristic of an emotional response.

(ii) The high road

The sensory thalamus stimulates the auditory cortex via appropriate neural circuits; the auditory cortex analyzes the location, frequency, intensity of noise, and so on, to determine the specific nature of potentially threatening auditory signals. In light of this analysis, the auditory cortex either inhibits or continues to stimulate the amygdala (and therefore the range of subcortical structures that elicit the effects characteristic of the emotional response), that is, in light of whether stimulus indicates neutral, positive, or dangerous events (Le Doux 1998, 2000).

2.2.2 Higher Cognitive Emotions or Higher Sentiments

Eighty percent of the human brain is cortex, and along with enabling toolmaking, language, and planning, the cortex enables the higher cognitive emotions (Panksepp 1998, 2000), that is, the emotional-affective processes that emerge from the recent evolutionary expansion of forebrain. These processes are responsible for human beings' subtle social emotions: shame, guilt, contempt, envy, humor, empathy, sympathy, and jealousy.

Unlike the basic emotions, higher cognitive emotions (HCEs) are traditionally considered not to display stereotypical patterns of physiological effects. They are more integrated with cognitive activity leading to planned, long-term actions than BE responses (Griffiths 1997, 100). The HCEs rely on symbolic representations that enable higher cognitive functioning, such as inferencing, reasoning, and planning. Human beings are uniquely endowed (some domesticated primates excepted) with symbolic representations that provide them with a completely internal locus of control.

Symbolic representations enable language, a social life governed by convention, and rich cultural traditions. The HCEs presuppose language, since they can be experienced only by creatures with the ability to semantically conceptualize; for example, shame minimally presupposes the conceptual grasp of self and responsibility for transgression. In addition, the cultural traditions that structure human social life account for differences in their fine-grained natures and the relative plasticity of their expression. What HCEs an agent experiences, therefore, are a function of her language and culture.

It is worth noting that higher brain functions are not essential for generating feelings. There is evidence that direct neocortical stimulation fails to generate affective states, but the evidence that various brain stem areas mediate affect is substantial. Damage to cortex only modulates the degree of emotionality, not the ability to have emotional feelings (Le Doux 1998). Clearly, what the cortex allows is ever-more-sophisticated ways for organisms to regulate their emotions (and, in human beings, how this ability develops, and the neurogenesis on which it is based, will be described in subsequent chapters).

2.3 The Functions of Emotions

Emotions are considered to have at least two functions: first, to monitor the environment and orchestrate the most appropriate response; and second, to enable the development and maintenance of social relationships. In terms of monitoring the environment and orchestrating the most appropriate response, emotions bring an animal's attention to relevant environmental properties and mobilize resources to deal with them (Frijda 1994; Scherer 1994):

- Some neural circuits are hardwired to solve specific adaptive problems (Panksepp 1998, 2000; Cosmides and Tooby 2000) and, when simultaneously activated, could deliver outputs that conflict, for instance, sleep and flight from a predator. To prevent such conflicts and maximize chances of survival, superordinate programs are developed that override some programs when others are activated.
- Some adaptive problems are best solved by simultaneous activation of many different neural circuits; for example, predator avoidance may

require simultaneous shifts in both heart rate and auditory acuity. Superordinate programs are required to activate and coordinate these differing circuits.

Emotions provide such programs; they are adaptations that have arisen in response to the adaptive problem of mechanism orchestration (Cosmides and Tooby 2000). This (probably) obtains for all vertebrate species.

More important for this monograph, however, is the function of emotion in enabling the development and maintenance of social relationships:

- Many emotions produce characteristic species-typical displays that broadcast to conspecifics the emotional state of the signaler (e.g., "I am afraid") and the identity of the recurrent evolutionary situation being confronted (e.g., "danger, above, in the local environment"). Many emotional expressions appear designed to be informative, and these have been so reliably informative that conspecifics have coevolved the appropriate response programs.

Vervet monkeys, for instance, use three distinct vocal-gestural alarm signals when threatened by predators. One alarm signals "snake," another signals "eagle," and a third signals "leopard." These different vocal-gestural alarms elicit the appropriate defensive responses in conspecifics, either to take to the trees or to take to the undergrowth. In addition, human beings have coevolved automated "interpreter" circuits to decode public facial displays of emotion into knowledge of others' mental states (Ekman 1984; Darwin 1872/1965), and when they cannot or will not verbalize intense emotion, they elicit from others typical comforting behaviors (e.g., hugging, kissing). It is also the case that smiling is a universally recognizable signal of readiness for friendly interactions (Izard and Ackerman 2000, 258).

- Facial expressions are required for normal attachment in infants and for attachment in courtship. Infants born with congenital facial paralysis have enormous difficulties forming any sort of social relationship. Facial expressions are also required to regulate aggression (Ekman 1984).
- People with emotional deficits due to damage to prefrontal cortex, for example, Phineas Gage, have considerable difficulty in making plans that involve others (Damasio 1994).

- Emotions seem to require expression (see earlier) and, in human beings, physiological and behavioral reactions are supplemented by linguistic expression. The urge to verbally share emotional experiences appears to be a cross-cultural phenomenon and affects both women and men equally (although what, how, and with whom they share may differ). Sharing initially takes place with intimates, but within just a few hours, through "secondary propagation," 50 to 60 percent of the relevant social group has been informed. This facilitates intragroup interaction and allows the group to keep track of the emotional status of members (Rime, Corsini, and Herbette 2002).

2.4 Feeling Theories of Emotion

"Folk psychology, our commonsense conception of the mind, tends to privilege conscious feelings. We ordinarily regard feelings as the essence of emotions. If a state did not feel like anything, it would not be a(n emotional) state" (Prinz 2004, 4). Another aspect of folk psychology, following Darwin (1872/1965), construes physiological changes as consequences of emotion: we feel fear, and our respiratory and heart rates increase dramatically, our musculature (including facial musculature) contracts in stereotypical patterns, and so on. However, William James (1884) famously challenges this thesis in his landmark paper "What Is an Emotion?":

> Our natural way of thinking about these standard emotions is that the mental perception of some fact excites the mental affection called the emotion, and that this latter state of mind gives rise to bodily expression. My thesis, on the contrary, is that *the bodily changes follow directly the perception of the exciting fact, and our feeling of the same changes as they occur is the emotion.* … Without the bodily states following on the perception, the latter would be purely cognitive in form, pale, colourless, destitute of emotional warmth. We might then see the bear and judge it best to run, receive the insult and deem it right to strike, but we would not actually *feel* afraid or angry. (13; James's italics)

Carl Lange (in 1885) concluded similarly, although he emphasized the perception of vasomotor changes, in particular, as implicated in emotional experience. More recently, David Papanicolaou (1989) and Antonio Damasio (1994) have resuscitated the feeling theory, although Damasio has modified it in certain respects. First, he includes the perception of neurochemical, including hormonal, changes in emotion response. Second, he suggests an "as-if loop" whereby the brain centers normally associated with bodily changes can be active in the absence of bodily changes. Third, Damasio argues that emotional responses can be perceived unconsciously (this is why his theory is somatic rather than somatic-feeling like James's and Lange's).

James uses three arguments to support his thesis. The first is the subtraction argument (also used by Lange [1885/1922]):

> If we fancy some strong emotion, and then try to abstract from our consciousness of it all the feelings of its characteristic bodily symptoms, we find we have nothing left behind, no "mind-stuff" out of which the emotion can be constituted, and that a cold and neutral state of intellectual perception is all that remains. (17)

To James, the phenomenology of emotional experience is exhausted by feelings of physiological perturbation: such perturbation, therefore, is the cause and not the effect of emotion. It is worth noting that his theory has found support in recent functional neuroimaging studies. It now seems incontrovertible that somatic brain centers (i.e., the centers in the brain that detect bodily changes) are activated during emotion induction (Damasio et al. 2000).

The second argument is the expression argument: "Everyone knows how panic is increased by flight, and how the giving way to the symptoms of grief or anger increases these passions themselves. ... Refuse to express a passion, and it dies" (21).

The third argument refers to the effects of pathology or lesions on emotion. James hypothesized that victims of total body anesthesia would experience seriously reduced emotionality (and one French physician at the time confirmed that he had a patient with precisely this affliction and that James was correct). Empirical evidence to support James's hypothesis was forthcoming in 1966, when Hohmann investigated patients with spinal cord lesions. He found that the higher in the cord the lesion was located, the greater was the patient's reduction in emotionality (sadness excepted).[4]

2.4.1 Objections to Feeling Theories of Emotion

(i) Readout of physiological changes is insufficiently fine grained to enable the distinction between subtle forms of emotions. The hair stands on the back of the neck in excitement in one situation and in fear in another. We feel the hair standing on the back of the neck in both situations. In addition, feeling in emotion individuation is insufficient (even if it is necessary) because some felt body states, such as fatigue and hunger, are not associated at all with emotionality.

(ii) The feeling theory fails to recognize the apparently frequent inclusion of thought in emotional experience. It just does not seem to be the case that we experience most of our feelings thoughtlessly. We can "think" ourselves into an emotional experience, we can think about what is going on while it is going on, and we can continue to think about it on its conclusion.

(iii) Some emotions, including fear and anger, require beliefs and judgments to elicit them. For example, blue exam books (Prinz 2004, 21) are not fearful in and of themselves (as coiled snake-shaped objects typically are); they only become fearful objects when linked to beliefs and judgments about what they are for, that is, to test or examine what has been learned, and how this can influence future career, prospects, self-esteem, and so on.

2.5 Cognitive Theories of Emotion

The locus classicus of cognitive theories of emotion is Kenny's (1963) *Action, Emotion, and Will,* which argues for conceptual links between emotion, propositional attitude ascription, and rational action. According to Kenny, any physiological state that lacks a suitable intentional object or fails to produce appropriate intentional behavior cannot be an emotion. It is logically absurd to insist that people can be afraid, for example, without believing that they are in danger or exhibiting avoidance behavior. The links between emotion, intentionality, and rational action, therefore, are conceptual and criterial (Griffiths 1997, 22), whereas the links between emotion and physiological states are merely contingent and empirical. This implies that only the logical links between emotion and belief and desire can form part of the definition of emotion; the contingent, empirical features of emotion cannot be part of its definition. This is because there is no contradiction in supposing an emotion to lack these physiological accompaniments.

An emotion has two objects, namely, particular and formal. A formal object is the property in virtue of which an event elicits an emotion, and a particular object is the event itself. The death of a child, for example, can be a particular object of one's sadness, but it causes sadness in virtue of being a loss. Being a loss is the formal object of sadness. Emotions represent their formal objects, not their particular objects. An episode of sadness may concern a variety of distinct, particular objects, but the sadness represents loss in each case (Prinz 2004, 62).

As Lyons (1980, 33) puts it: "In general, a cognitivist theory of emotion is one that takes some aspect of thought, usually a belief, central to the concept of emotion and, at least in some cognitive theories, essential to distinguishing emotions from one another." Cognitive theories of emotion have been termed propositional attitude theories by Griffiths (1997, 23). A propositional attitude is a mental state consisting of a representation of a proposition and an attitude (e.g., belief, desire, fear, hope, etc.) toward that proposition. A proposition, in turn, is a declarative statement that describes a state of affairs, such as "today is Tuesday." This state of affairs is the propositional object of the mental state, and it is preceded by a statement of the attitude directed toward it and the word *that*. Thus I believe *that* today is Tuesday, or I suppose or hope or fear *that* it is Tuesday (De Lancey 2002; Prinz 2004).

The foregoing account construes emotions as reducible to beliefs, but it would also hold for theorists who view emotions as reducible to desires (e.g., to flee or freeze). The important point is that the ascription of propositional attitudes to an emoting subject entails the ascription of the appropriate concepts to her. If I state that I am afraid that I will be attacked in the dark alley that lies ahead of me, we can assume that I possess concepts, for example, of fear, of being attacked, of danger, and of dark alleys. It is for this reason that Prinz (2004, 23) claims that cognitivist philosophers construe emotions as essentially conceptual. This is the basis of his conceptualization hypothesis claim. He also claims, however, that cognitivist philosophers are committed to two further hypotheses, that is, the disembodiment and appraisal hypotheses.

The disembodiment hypothesis (Prinz 2004, 25) denies that the cognitions that constitute our emotions are identical to somatic changes or to internal states that register these changes. Some cognitivists even deny that somatic changes are required at all in emotional experience (e.g., Solomon 1977; Nussbaum 2001), while others agree that such changes are necessary (e.g., Lyons 1980) but claim that nonsomatic cognitive states are necessary, too. They all insist, however, that the cognitions associated with emotions are something above and beyond somatic perturbation or the states that register these perturbations and, as previously indicated, that such cognitions individuate the emotions (Gordon 1987; Lyons 1980). The cognitive components bound to our emotions are, in this account, disembodied (Prinz 2004, 25).

The third hypothesis underpinning cognitivist philosophies of emotion is that of appraisal. Appraisals are representations of an organism-environment relationship that bear on well-being. It is interesting to note, along with Prinz (2004), that despite apparent dissimilarities, the theories of cognitivist philosophers and psychologists are extremely similar. Psychologists prefer to talk of appraisals rather than propositional attitudes; appraisals, however, take the form of judgments (Solomon 1977; Lazarus 1991). Anger, for example, comprises the judgment that there has been an offense against me or the people I love or admire (and judgments are quintessentially appraisals). Forming the judgment that there has been an offense, in addition, presupposes the concept of offense. A second point of similarity is that cognitions are necessary to individuate emotions. Psychologists differ only insofar as they construe emotion as being caused by appraisals rather than as identical to or constituted by them. A third point of similarity is the construal of emotions as highly structured. Anger, for instance, involves more than the simple judgment that an offense has been committed. The highly structured nature of emotions, according to psychologists of emotion, is captured in their dimensional appraisal theories.

2.5.1 Dimensional Appraisal Theories

To appraise something is to judge it as somehow salient to one's well-being (Arnold 1960; Lazarus 1991; Frijda 1986), and emotions, according to appraisal theorists, include appraisal judgments. Dimensional appraisal theorists claim, in addition, that all such appraisal judgments are drawn from a common set of appraisal dimensions. According to Arnold (1960), there are three dimensions involved, namely, whether the situation faced is beneficial or harmful, whether it involves objects absent or present, and whether those objects are easy or difficult to avoid. Different "answers" to dimensional appraisal "questions" elicit different emotions. Joy, for example, involves objects that are beneficial, present, and easy to obtain.

Objections indicating the inadequacy of these three appraisal dimensions to distinguish emotions (e.g., fear and anger both involve objects that are harmful, present, and difficult to avoid) have resulted in (at least) one hugely influential multidimensional appraisal theory, by Lazarus (1991). This appraisal theory includes six dimensions that are classified as primary or secondary appraisals. Primary appraisals establish that something salient to well-being is or could be present in the organism-environment relationship, and secondary appraisals relate to the adequacy of the emoter's coping resources.

Primary appraisals refer to goal relevance (is a given transaction with an object, event, or situation relevant to one's goals?), goal congruence (does the experience of the object, event, or situation facilitate or impede the fulfillment of one's goals?), and type of ego involvement (what in terms of self-identity, moral values, self-esteem, life goals, etc., is at stake here?). Again, answers to these dimensional appraisal questions determine the positivity or negativity of the emotional response.

Secondary appraisals include the ascription of blame or credit (who or what is accountable for the object, event, or situation, and should blame or credit be assigned?), the estimation of coping potential (what resources are available to deal with the event, object, or situation?), and future expectancy (will things change to become more or less congruent with my goals?).

Lazarus further classifies his appraisals as either molecular or molar. Molecular appraisals are the products of his six appraisal dimensions; that is, they correspond to the judgments that people make before eliciting an emotional response. The combination of molecular judgments associated with anger, for instance, lead to the recognition that one has been offended against. This superordinate judgment is the molar appraisal. Molar appraisals, in turn, capture what Lazarus famously construes as "core relational themes," which are relations that bear on well-being. Lazarus (1991, 122, table 3.4) identifies fifteen such core relational themes associated with distinct emotions. These emotions are anger, anxiety, fright, guilt, shame, sadness, envy, jealousy, disgust, happiness, pride, relief, hope, love, and compassion. Core relational themes (CRTs) are not explicitly represented attitudes; rather, they represent the basic situations that emotions are designed to discriminate. For instance, the CRT associated with anger is a demeaning offense against me and mine; sadness is associated with the CRT of irrevocable loss; and fright is associated with the CRT of immediate, concrete, and overwhelming physical danger.

In sum, appraisal theorists construe emotions as preceded by conceptual appraisals and conceptual appraisals as dimensionalized to a greater or lesser extent.

Appraisal theorists (obviously) espouse the propositional attitude hypothesis of philosophers; it is clear, in addition, that they share conceptualization and disembodiment hypotheses, too. In general, therefore, such theorists are vulnerable to the same objections as philosophers. In addition, Lazarus and other "goal" theorists (e.g., Oatley 2004) are vulnerable to another objection: not all emotions are related to goals. I can be delighted by a surprise gift without my delight being in any way goal related. Similarly, I can be disgusted, either physically or morally, without the object being related to goals (unless one insists that uncontaminated food and moral "goods" are goal related in some way, but this does seem rather desperate).

2.5.2 Objections to Cognitivist Theories of Emotion

(i) Emotions are underdetermined by beliefs and judgments (Zinck and Newen 2008; Prinz 2004; De Lancey 2002; Griffiths 1997); that is, even if they were necessary (which they are not), they are not sufficient to individuate emotions. By reducing emotions to propositional attitudes, cognitive theories fail to distinguish emotions from them (Goldie 2000; Zinck and Newen 2008). Reductive cognitivism fails on empirical grounds, too, for if emotions reduce to judgments, then the relevant judgments should always be accompanied by the proper emotion. Research, however, demonstrates that they are not (Damasio, Tranel, and Damasio 1990). Patients with damage to areas of prefrontal cortex show no loss of cognitive capability but manifest obvious emotional defects.

(ii) Emotions can be elicited directly by electrical stimulation of relevant brain regions, either by electrode or even by epileptic stimulation (Frijda 1986; Panksepp 1998) and by hypothermia, fatigue, and vigorous exercise (De Lancey 2002). In addition, at least one emotion, anger, can be triggered by faulty cooling of the brain, which can result from sinusitis (Zajonc, Murphy, and Inglehart 1989).

(iii) Emotions can be elicited by voluntary changes in the body, especially by changing its musculoskeletal orientation and facial expression (Ekman 1972; Lange 1885/1922).

(iv) Emotional responses are elicited by perceived objects, that is, by concreta (De Lancey 2002), for example, parental faces and snake shapes in human infants and predators in nonhuman infants, neither of which can semantically conceptualize. To deny that such responses are emotional in the face of overwhelming empirical evidence (e.g., Panksepp 1998), as, for example, Ortony, Clore, and Collins (1988) do, seems extraordinary.

(v) Emotional responses can be triggered by imagination. People can experience full-blown emotions by simply imagining suitable objects (Greenspan 2004) or even while "enjoying" what they clearly understand to be fictional events and situations. Beliefs and desires, in these situations, are ones the emoter explicitly does not possess (Griffiths 1997, 29).

(vi) Emotional responses are characterized by a whole suite of physiological changes. These include changes in facial expression (De Lancey 2002; Ekman 1984, 2004), musculoskeletal function (flinching, expressive vocal changes), and autonomic nervous system changes (adrenaline release, changes in heart rate, respiratory rate, etc.). These physiological changes form stereotypical patterns of responses associated with particular emotions (De Lancey 2002; Levinson 2006). Importantly, too, recognizing emotional responses in others depends on these factors (Ekman and Friesen 1969). Cognitivist theories cannot explain these phenomena.

(vii) Emotions or emotionlike responses are present in many nonhuman animals. If we accept evolutionary theory, then we should expect there to be homologues of many emotional capabilities in many nonhuman species, with more closely related species sharing more common features. We commonly talk of dogs and cats experiencing fear, but we never attribute this emotion to worms. If we deny that such responses are emotional (e.g., Ortony, Clore, and Collins 1988), then we have to acknowledge that scientists who attempt to study emotion, especially fear, in nonhuman animals are seriously mistaken. It is also worth noting that when the use of emotion terms is prohibited in the description of primate behavior, the objective behavior statements are inadequate to characterize behavior or to identify individual animals and safely predict their behavior (Hebb 1946). Scientists are left with no explanation of the animal behaviors in question.

(viii) Burgeoning neuroscientific evidence suggests that at least certain emotions are associated with certain brain areas and neural circuits. Fear, for instance, is associated with the amygdala (Panksepp 1998).

(ix) Emotional "objects" can be displaced. My anger that my employer underestimates my capabilities can be displaced onto the cafeteria assistant who offers me cold tea (De Lancey 2002, 41). If cognitivism were true, then these two judgments would elicit two different emotions; they do not.

(x) Human beings show evidence of the development of emotional capabilities from early infancy right through to mature adulthood (Ekman 1984), and infants develop affective capabilities before cognitive capabilities (De Lancey 2002). Neonates show evidence of the facial expression of emotion and, after only a few weeks, exhibit the behavioral manifestations of basic emotions, such as anger at being frustrated and joy or happiness in the presence of the primary caregiver's moving face. Cognitive theories fail to explain these phenomena, too.

2.6 The Social Construction of the Emotions

The foregoing account, especially as it relates specifically to basic emotions, demonstrates that some theorists (e.g., Ekman 1984, 2004) are convinced that some emotions, or at least their expression, are pancultural phenomena. Others, however, deny that emotions can be pancultural, and cite a range of putatively idiosyncratic emotions or emotion analogues to support their thesis. Such theorists are social constructionists.

Social constructionism can be weak or strong: all social constructionists, however, believe that emotions are, in some sense, created by culture. Indeed, strong social constructionists (e.g., Lutz 1986; Harre 1986; Averill 1980) insist that emotions are only constructs of culture and possess no pancultural features of the kind that biology or any other physical science might properly describe (De Lancey 2002, 71). More precisely, to claim that an emotion is socially constructed is to claim that its identity conditions depend on features that are a function of the sociocultural environment in which the subject emoter is embedded. "The identity conditions of such an emotion advert to a property that emerged through the influence of beliefs, values, practices and institutions that are specific to a particular human group. Culture makes a constitutive contribution" (Prinz 2004, 131–132).

Some of the emotions that social constructionists claim are peculiar to certain human groups and which are alien to Anglo-American culture include the following:

Gezellig: a Dutch emotion (or emotion term) that means something like feeling cozy in English but occurs only in the presence of other people (Harre 1986).

Amae: a Japanese emotion or emotion term that describes a feeling of pleasant dependence on individuals and institutions (Morsbach and Tyler 1986).

Song: an Ifaluk emotion or emotion term that describes an analogue of anger (Lutz 1986) but is more akin to simple admonition with moralistic overtones with no disposition to revenge.

In addition, the phenomenon of romantic love is peculiar to Western cultures (Griffiths 1997; Averill 1980), and emotions, or their expression, are influenced by the individualistic or collectivist nature of their host culture (Prinz 2004; De Lancey 2002), which I discuss in detail in chapter 4.

2.6.1 Construals of Social Constructionism

There are three different construals of social constructionism:

(i) A trivial sense. All concepts are socially constructed; no concept can exist independently of a community of speakers, and each concept is created by sociolinguistic processes.

(ii) A strong sense. Citizens, members of Parliament, and husbands are social constructions in a way that gold, electrons, and hormones are not. The latter phenomena would exist, and examples of them would have certain properties in common, irrespective of the social construction of their conceptual categories. The category of member of Parliament, however, depends on the formulation of the concept of MP. The sociolinguistic processes focused on this concept imbue MPs with properties that differentiate them from non-MPs.

(iii) A very strong sense. A concept exists, but that to which it putatively refers does not; the important point, however, is that people who use the concept fail to recognize this. Young men of the Gururumba, for example, manifest a "wild pig" syndrome, and the Gururumba people, in general, accept that this syndrome is real. Such practices or pretenses (Griffiths 1997, 147) cannot survive the realization that they are mere thoughts. They must be treated as natural by the relevant community.

Very strong social constructionists (VSSCs) like Harre (1986) and Lutz (1986) believe that emotions are just like the Gururumba's wild pig syndrome:

> Psychologists have always had to struggle against a persistent illusion that in such studies as those of the emotions there is something there, the emotion, of which the emotion word is a mere representation. This ontological illusion, that there is an abstract and detachable "it" upon which the research can be directed, probably lies behind the defectiveness of much emotion research. (Harre 1986, 4)

2.6.2 Objections to Very Strong Social Constructionism

Very strong social constructionism, at least as it relates to the emotions, is untenable. As a view, it is both conceptually incoherent and evidentially flawed. In terms of conceptual incoherence, the denial of any emotional panculturalism is lethal. If no common identity conditions exist for different emotions, then ethnographers and anthropologists have no way to identify emotions or emotion analogues in other cultures. Indeed, we would have no reason to assume that other cultures experience emotions at all.

There are a number of other objections to VSSC:

(i) It seems unwarranted to claim that different cultures have different emotions when, at best, research demonstrates, rather more modestly, that different cultures manifest different eliciting conditions and response repertoires.

(ii) It also seems unwarranted to claim that emotions are social scripts when the research suggests, again rather more modestly, that they might have scripts as components. Particular emotions can possess a number of scripts; romantic love, for instance, has both a "falling in love at first sight" script and a "growing to love another" script. It also has a "sex expectancy after marriage" script in certain cultures, and a "no sex expectancy" script in at least one other culture (in India).

(iii) There seems little principled reason to assume that differences in vocabulary point to the social construction of emotions (Pinker 1994). It is not implausible that non-Dutch and non-Japanese people could experience something closely akin to *gezelligheid* and *amae* even though they possess no emotion term for this feeling. *Gezelligheid* and *amae* might easily be felt by two close friends enjoying a beer together in their favorite English public house (Prinz 2004, 136).

(iv) There are indications that VSSCs fail to distinguish formal from particular objects of emotion. For instance,

witiko among the Algonquin Indians is a morbid fear of having turned into a cannibal with a craving for human flesh (Trimble et al. 1984);

p'a-leng in China is a morbid fear of the cold (Kleinman 1980); and

koro in South China refers to an intense anxiety that penis or vulva and breasts will retract back into the body (Yap 1965).

These terms are all directed at different particular objects (cannibalism, cold, genital retraction), but all have the same formal object, namely, danger or threat. This suggests that they are all fears. This, in turn, suggests that the fact that people of different cultures fear different objects does not count as evidence for different emotions across cultures.

(v) Finally, and pace Harre (1986), when we speak of emotions, we are speaking of something that *is there*. Burgeoning evidence from neuroscientific research suggests that some emotional capabilities are significantly biologically determined (Panksepp 1998; le Doux 1998); at least some pancultural emotional facial expressions exist (Ekman 1984); other nonhuman species possess emotions (Darwin 1872/1965; Hebb 1946); and certain neural circuits are associated with certain emotions (Panksepp 2000; Damasio 1994; Le Doux 1998). Smiling, for instance, is not just a socially learned behavior that is part of our socially constructed emotion of joy. An independent neural pathway for facial control enables spontaneous smiling in hemiplegics, and congenitally blind-deaf-dumb and brain-damaged children (Eibl-Eibesfeldt 1973) spontaneously laugh and smile when playing but cry and shout when put into unfamiliar situations.

2.7 More Recent Theories of Emotion

Three more recent theories of emotion deserve brief attention in this review. These are Zinck and Newen's (2008) multifactorial, developmental account, J. Clark's (2010) serial homology account, and Colombetti's (2014) enactivist account.

2.7.1 Zinck and Newen

My own account of human emotion is consistent with that of Zinck and Newen (2008) in that (i) the approach adopted is developmental and functional, and (ii) emotions are essentially expressive devices (this is unremarkable given that we support our argumentation using the same developmental psychological research). Where our approaches differ, however, is in our construals of precursor emotions. Zinck and Newen identify only two prebasic emotions, comfort and distress, which they distinguish from felt body states like homeostatic distress, for example, hunger. They distinguish them on the grounds that they do not evolve into more distinguished behavioral patterns as the "complex class of emotions" does (8). I argue that the unfocused crying, motor unrest, and cooing and snuggling are the only expressive responses neonates possess with which to respond to the distinct interoceptive and proprioceptive stimulation associated with precursor emotions and triggered by homeostatic distress. In addition, I argue that the developmental mechanisms (which Zinck and Newen fail to address) from emotion precursors (affect expressions) to basic to higher cognitive emotions are the same. In addition, it has recently been argued that elements of more primitive types of emotion persist throughout phylogeny. This is the view of J. Clark.

2.7.2 J. Clark

Clark's (2010) serial homology account is another important recent theory. As previously noted, basic emotions are considered to possess a number of features that distinguish them from HCEs (Ekman 1992). These are characterized as involving relatively stereotypical physiological, behavioral, and expressive patterns that are underpinned by specialized neural circuits, and as being genetically hardwired as a result of a long evolutionary history. BEs can be activated by unconditioned stimuli and manifest a degree of modularity such that neither the unconditioned stimuli that elicit them nor the responses that are automatically generated can be moderated by higher cognitive functioning. These stimuli, it is claimed, have a privileged strength that makes their "unlearning" impossible or very difficult indeed. However, despite possessing relatively simple unconditioned elicitors, BEs can also be activated by conditioned stimuli through learning. Finally, BEs are considered to develop early in life, before the emergence of higher cognitive capabilities. In contrast, HCEs are thought to lack many of these features (Griffiths 1997). They are typically construed as uniquely human and thus lack a shared evolutionary history with other animals. They are also considered to lack stereotypical physiological, behavioral, and expressive features. HCEs are considered to be variable across cultures, and any characteristics they possess are learned through and relative to the embedding culture. While both BEs and HCEs are considered capable of responding to higher cognitive input, HCEs are considered to lack simple unconditioned elicitors. Both the input and output sides of HCEs, therefore, are viewed as much more flexible than BEs.

Using shame as his example, Clark (2010) persuasively argues that BE shame and HCE shame are serial homologues. Homologues are traits that organisms share in common as a result of descent from a common ancestor. Homologous traits must all occur within a continuous line of descent but can cut across species and functional lines; for instance, dolphin flippers and human hands are homologues (78).[5]

Clark argues that many features and functions of BE shame are directly retained in human forms of shame, to the extent that dominance hierarchies still play a role in human life (81). Shame in humans emerged from a rank-related emotion in nonhuman mammals, the primary function of which is to signal subordination to a dominant conspecific so as to appease him or her. The functional breadth of shame in humans has expanded in at least two ways. First, while the dynamics of dominance hierarchies continue to play a role in human social hierarchies, the focus has largely shifted from dominance (coercive power) to "prestige competitions" involving the attraction of positive social attention and the possession of socially valued traits or materials. Shame has come to be associated with loss in prestige competitions. Second, shame has acquired the function of facilitating cooperation and conformity by signaling one's own awareness of one's violations of social norms, as well as one's desire to conform to them. This signals our reliability as partners for cooperative activities, which facilitates cooperation with others and enhances social reincorporation when these norms have been violated:

> Despite the expanded functions of shame in humans, its physiological, expressive behavioural and perhaps some affective cognitive effects have been largely conserved. For example, individuals in situations involving prestige or cooperations/conformity shame still display the expressive and behavioural displays characteristic of BE shame. (J. Clark 2010, 81)

2.7.3 Colombetti

Giovanna Colombetti (2007, 2009a, 2009b, 2010, 2011, 2014; Colombetti and Krueger 2015) is a leading contemporary theorist in extended or, probably more accurately, extensive cognition and affectivity (see Hutto and Myin 2013; Hutto, Kirchhoff, and Myin 2014), whose approach is that of enactivism. Enactivism, as she points out, is a synthesis of several interrelated and mutually reinforcing insights from a variety of different disciplines, especially phenomenological philosophy, psychology, biology, and neuroscience. Enactivism denies that cognition is realized centrally by the brain; rather, it claims that cognition is realized (enacted) by the whole living organism, deeply embedded in the world, and because of this, putative boundaries between brain, body, and environment are also denied. This denial distinguishes extensive from extended affectivity. Affectivity is defined as a lack of indifference, or a sensitivity or interest in one's existence, and both cognition and emotion are instances of "the relentless sense-making activity of the precarious living organism as it maintains itself via continuous processes of self-regulation and exchange with the environment" (Colombetti 2014, xviii). Enactivism views life as sufficient for cognition and sees all living systems as affective and cognitive (18). The fundamental life regulatory processes that enable organismic survival are already, in a primordial sense, meaning generating and thus affective.

A central feature of Colombetti's enactivism (and enactivism in general) is its focus on the lived experience; enactivism adopts a distinctly phenomenological perspective, due primarily to Husserl (1952/1989) and Merleau-Ponty (1962), in that the body is construed not merely as a physical entity but also as an experienced, lived structure. Accordingly, comprehending mentality and emotionality requires careful exploration of embodied nature at both the physical and experiential levels. These explorations should be mutually informative; descriptions of lived experience are required to make sense of brain and bodily activity, and data from brain and bodily activity should be used to refine experiential reports.

As an enactivist, Colombetti views emotional episodes, such as anger, and happiness, as self-organizing patterns of the entire organism that recruit neural, muscular, autonomic, and other processes into highly integrated configurations (2009a, 2014). Self-organizing emotional episodes can be highly variable because the processes constituting them can self-organize in different context-dependent ways. Importantly, however, the range of possible variations depends on the state of the organism and is evolutionarily and developmentally constrained. The enactivist perspective construes "internal causes" of emotion such as affect programs and the concept of basic emotions as redundant; emotional episodes are construed as complex, flexible, and variable self-organizing patterns (see also M. Lewis 2005; M. Lewis and Granic 2000).

The views of Colombetti, especially as they relate to extended (but not extensive) affectivity, the role of language in emotional experience and expression (see her 2009b), her emphasis on operational constraints and context dependence, and her strong advocacy for a multidisciplinary approach to affective science, are broadly consistent with mine; our views differ, however, in a number of significant respects.

First, I view affectivity as evolving to enable and regulate the lives of social species; I agree that even the simplest of organisms possess regulatory systems (e.g., chemotactic systems), but affectivity proper is, in my view, restricted to social animals. Second, while agreeing that affectivity is experienced by whole organisms, I disagree that it is a whole-organism response (although how much our disagreement is a matter of emphasis is a moot point). Complex creatures like mammals, especially human mammals, have evolved particular systems, which I term "realizer systems" in chapter 3, to enable complex functionality. In chapter 3, I describe the total realizer systems of human emotionality, together with their partial and core realizers and the background conditions that sustain them. Claims relating to whole-organism response obviously mask these distinctions; indeed, such claims conflate realizer systems and background conditions. Emotions might be experienced by whole animals, but they are realized subpersonally by particular realizer systems. Third, I do not agree that affectivity and cognition are always unbounded in the sense she describes; if no boundaries exist between brain, body, and world, then the functional contribution of individual components to a total realizer system cannot be identified. This would seem to hinder rather than to assist the detailed description and explanation of contributory phenomena. A related point, which I argue in later chapters, is that affectivity and cognition may well be unbounded in the sense she describes when they are up and running, but interaction is initiated typically by one partner and terminated by one partner or by either explicit or tacit agreement among partners. At the point of initiation and termination, at least, there are thus identifiable loci of interactive control.

Two further points of difference are worth mentioning. The first is that Colombetti sees affectivity as being sense making, whereas I see it as being specifically salience detecting and salience expressive (although, again, it is a moot point how far our construals here differ). In addition, our views relating to affect programs and the basicality of emotion differ (see chap. 4), and my theorizing is restricted to extended rather than extensive affectivity. The explanatory framework of deep functional integration that I favor is adequate to the description and explanation of the data I review.

There is one additional influence on human affectivity that both Colombetti and I recognize as central, but here again our emphases differ. Colombetti tends to focus on adult affectivity, whereas I emphasize emotional development. The influence in question is scaffolding.

2.8 Scaffolding of Emotional Development

A recent apparent burgeoning of interest in the scaffolding of affectivity, including developmental affectivity (e.g., Colombetti and Krueger 2015; Krueger 2013; Greenwood 2013; Sterelny 2010), that appears connected to various species of situated cognition (see chap. 3) is an emotion-focused offspring of Vygotsky's (1978) more general sociocultural theory of development. Vygotsky's theory includes the concept of the zone of proximal development (ZPD), which represents what a child can accomplish with the guidance and support of an adult compared with what she can accomplish independently. Vygotsky suggests that a child's future independent performance largely depends on the guidance provided by her caregiver(s) in the ZPD. The caregiver creates and mediates the learning process by providing culturally appropriate guidance. This theory informed the work of Wood, Bruner, and Ross (1976), who introduced the metaphor of scaffolding, derived from the construction industry, to describe teachers' scaffolding of student learning in a puzzle task. It has been used most frequently, however, to describe the kinds of instructional exchanges that take place in informal educational situations, such as parent-child interactions. The concept was developed further by Rogoff and Wertsch (1984), who introduced the notions of orchestration and transfer to describe the process of development from other-regulation to self-regulation. Successful scaffolding requires establishing a shared understanding of the task (Rogoff 1990), or "intersubjectivity." In this process, the adult scaffolds the development of the child's own conception of the task by providing appropriate guidance and support.

Two integrated functions operate in the mother-child interactions that promote gradual self-regulation in children. These are (i) the mother's sensitivity to her child's level of performance and provision of appropriate guidance, and (ii) the mother's gradual withdrawal of support, which allows the child more opportunity to perform independently (Rogoff 1990; Winsler, Diaz, and Montero 1997). The mother initially assumes responsibility for regulating and managing the activity, but as the child becomes more competent, the mother gradually transfers the responsibility to the child. Scaffolding thus includes the notion of reciprocal exchange, where participants adjust the way in which they understand something in relation to each other (Valsiner 1984). Scaffolding is a process that changes both the content and the structure of the interactions between the person doing the scaffolding and the person being scaffolded in relation to the strengths and needs of the person who is being scaffolded (Pea 2004). It is an important and seemingly inevitable process in the normal development of children's self-regulation (see chaps. 4 and 5).

In terms of the development of self-regulation, scaffolding refers to an adult's control over elements of a task initially beyond the capabilities of the child to enable the child to concentrate her efforts on elements that she is currently capable of managing (R. Clark, Menna, and Manel 2013). As scaffolding progresses, the adult remains sensitive to the child's successes and failures, instituting further instruction according to the child's performance (Wood, Bruner, and Ross 1976). Effective maternal scaffolding includes verbal scaffolding, autonomy support, sensitivity, and contingent responsiveness to infant performance, all of which are included in intuitive parenting skills.

Research demonstrates that when scaffolding is provided in a developmentally sensitive fashion, it can, for example, reduce the development of aggressive behavior in preschoolers (R. Clark, Menna, and Manel 2013); improve executive functioning in children born with very low birth weight and normal weight (Lowe et al. 2014); promote attention regulation in preschoolers from low socioeconomic groups (Robinson, Burns, and Davis

2009); and facilitate the development of IQs over 130 at four years (Morelock, Brown, and Morrissey 2003).

Despite the recent nature of some of this research, and the recognition that sensitivity and responsiveness to the infant's needs are crucial to development, some researchers still tend to focus unidirectionally (see chap. 1), that is, on the contribution of the mother or caregiver rather than the contribution of the child to the scaffolding process (Mascolo 2005). In addition, such research fails to identify what it is that the child must do (contribute) to benefit from scaffolding interactions with others (Mascolo and Fischer 2005; Rogoff 1990). A further weakness (Butler 1998) is that a primarily unidirectional view of scaffolding suggests a relatively fixed and predetermined goal in the development of scaffolded activity. Recognition of these limitations, enabled by the invention of highly sophisticated micro-videotaping techniques, has led to the development of bidirectional, transactional, or circular models of scaffolded interaction (Maccoby 1992) and, more recently, to dynamical systems models (e.g., Fogel, King, and Schanker 2008; M. Lewis and Granic 2000) and coaction models (Gottlieb 2002; Mascolo 2005; Gottlieb and Halpern 2008).

From a coactive systems approach, the person-environment system functions as the primary unit of developmental analysis. This system involves at least five classes of coacting elements: "Within a given *socio-cultural context*, *individual actions* are directed toward some physical or psychological *object*. In interaction with *other persons*, individuals engage dialogically using a variety of different *mediational means*, including signs, symbols and other cultural tools" (Mascolo 2005, 186–187; his italics). In a coactive system, although these components are distinct from one another, they are inseparable as causes of action and development; control over the construction of action and meaning is distributed throughout the coacting elements of the person-environment system (Wertsch 2002; Granott 1998). "Although many approaches have embraced an interactive approach to development, empirical analyses often tend to focus on the independent contributions of one or more elements of the person-environment system. However, moment-to-moment analyses of changes in the person-environment system can illuminate how subtle and non-obvious coactions among system elements can create novel ways of thinking, acting and feeling" (Mascolo 2005, 186). Relational activity analysis provides a method for identifying the precise structure of joint action; it identifies which elements of the interaction are controlled by whom and also the coregulated scaffolding that occurs between actors. Coactive scaffolding occurs when elements of the person-environment system beyond the direct control of an individual direct or channel the construction of an action in novel or unanticipated ways. However, it should be acknowledged that novelty is not unrestricted. In G. Bateson's (1972) famous example of ecological scaffolding of tree chopping, features of each element (e.g., musculature of chopper, weight of ax, radius of tree, etc.) all function as part of a feedback system. Each element exerts a constraint on the chopping process and thus functions as an integral part of the tree-chopping action. Feedback continues throughout the chopping action, introducing new constraints and channels for action.

Three varieties of coactive scaffolding have been identified, namely, social, ecological, and self-scaffolding (and all three are exemplified in the development of human emotionality, language acquisition, and symbolic thought). Social scaffolding refers to the process by which coregulated exchanges with other persons direct development in constrained directions. Ecological scaffolding refers to the ways in which one's relation to, or position within, the broader physical and social ecology moves action toward novel forms. Any action necessarily occurs within the broader physical and sociocultural context that provides feedback to individual action or otherwise constrains and directs others (G. Bateson 1972; Wertsch 1998). Self-scaffolding refers to the ways in which an agent directly or indirectly changes the environment to suggest new meanings or cognitive operations, for instance, moving individual tiles to assist in the compilation of words in the game of Scrabble.

2.9 Basic Emotion and Emotion as Natural Kind

2.9.1 What Emotions Are Basic?

The debates concerning the nature and relationship of BEs and HCEs fuel two further debates concerning the emotions. These are the basicality or nonbasicality of the emotions and their natural-kind status. Theorists who favor a blending or subsumption of certain emotion categories, that is, the unity theory adherents, also tend to favor the basic–nonbasic distinction and the natural-kind status of the emotions (e.g., Prinz 2004). Such theorists consider that there is a privileged set of basic emotions from which others are derived. Those who consider BEs and HCEs too different and too diverse to blend or subsume, that is, the disunity theory adherents, tend to reject the basic–nonbasic distinction and claims to natural-kind status (e.g., Griffiths 1997). Such theorists deny that there is a privileged set of basic emotions from which others are derived.

There are three ways to characterize basicality:

- Psychologically primitive (Ortony and Turner 1990), which construes basic emotions as only those that are not derived in any way from other emotions.
- Biologically primitive (Ortony and Turner 1990), which construes basic emotions as being innate or present in all normally developing members of a species.
- Emotions for which dedicated neuroanatomical regions and neurochemicals can be identified (Panksepp 2000).

Panksepp's view is intuitively attractive in that dedicated neural circuitry could be considered as evidence of innate endowment and nonderivativeness. Unfortunately, however, one familiar feature of neural networks is that different states can supervene on distinct patterns of activation across the same population of neurons.

Theorists who favor the idea of basic emotions unanimously identify the affect programs as psychologically and biologically primitive.[6] Griffiths (1997) has been the most trenchant critic of the basicality thesis; he has vehemently denied that HCEs (e.g., guilt, jealousy) have BE components. As I have just noted, however, this position has been strongly challenged by J. Clark's (2010) theory of serial homology (and I have already noted, in chap. 1, Griffiths's [2003] and Griffiths and Scarantino's [2009] more recent concessions concerning the relationship of BEs to HCEs and how this might usefully be discussed within a common theoretical framework).

2.9.2 Are Emotions Natural Kinds?

The answer to this question will be affirmative if it can be shown that all emotions are or are derived from basic emotions. If emotions do not form a natural kind, as disunity theorists opine, there can be no single science of the emotions. Different explanatory frameworks will be required to explain different emotion categories.

A natural kind is generally construed (obviously) to be a "natural" phenomenon in that its boundaries derive from nature and not from the way human beings choose to classify such phenomena. Nature has its own inbuilt "joints," and it is the business of science to identify them (Prinz 2004, 80). For example, lions, tigers, bears, water, gold, and the planets, and so on and so forth, are natural kinds. Good manners, poor arguments, and beautiful horses are not.

Philosophers disagree on what precisely distinguishes natural kinds from nonnatural kinds. There are at least two opposing views. The first view is Kripke's (1980), who insists that members of a natural kind all share a common underlying essence. All dogs are dogs, in this account, in virtue of sharing a common canine genome; all samples of water are water in that they are composed of H_2O molecules. The underlying essence in Kripke's natural kinds, therefore, is a function of fixed properties. The underlying common-essence theory, however, is difficult to apply to certain other (natural) categories. Rivers, for instance, may not be composed of identical materials.

The second view is that of Boyd (1989, 1993), who has a more liberal construal of natural kinds. He includes homeostatic property cluster phenomena, which typify complex, relational, and functional kinds like biological, psychological, and social kinds. The properties of such kinds are never fixed; they can be lost or gained through a change in the kind's relations or in the context in which the kind exists or occurs. Biological kinds are essentially what they are in virtue of the interaction of morphology, physiology, and behavior (Boyd 1993). Species, for instance, are necessarily indeterminate; each species is an intermediate between its parent species and its emerging daughter species. The example of species is instructive; they typically show variation, not in some peripheral subset of their features but in all of them. No character, however central to our stereotype of that species, to its morphology, or to its genotype, is essential for an individual organism to be a member of the species. Tigers exist without stripes, calves are born with two heads, and Down syndrome sufferers are human despite an additional chromosome. Biological kinds, therefore, are homeostatic property cluster kinds. Homeostatic mechanisms bring about the co-occurrence of their properties in nature. Some properties of the cluster kind must be present for the cluster kind to be that kind, but not all the cluster properties must be present for the cluster kind to be that kind, unlike H_2O

and water (Greenwood and Bonner 2008). The account of human emotional ontogenesis provided in this monograph demonstrates clearly that emotions are homeostatic property cluster natural kinds.

BEs and HCEs form a natural kind because they are serial homologues. They are homologues in that their shared line of inheritance accounts for the various deep commonalities in their underlying mechanisms at various levels of organization, including genetic, ontogenetic, physiologic, behavioral, computational, and others (J. Clark 2010, 84–85). In addition, they are homologues when assessed against Remane's (1952) operational criteria for identifying and testing the homologues of higher-level traits.[7] These criteria are the following:

(i) *Position*: the spatiotemporal position or functional role of traits within a more general pattern of organization.
(ii) *Special quality*: the complexity, distinctiveness, or specialization of the traits (which makes it less likely that they evolved separately).
(iii) *Continuity*: the presence of an evolutionary continuum of properties, from a simpler and more productive state to a more complex and derived state of the trait.
(iv) *Shared developmental pathways*: the use of two common developmental mechanisms or constraints to establish the traits (J. Clark 2010, 84).

Clark's analyses relate to evolutionary time, of course, and he argues that both BE and HCE shame and pride are natural kinds. The account of human emotional ontogenesis I present in this monograph relates (of course) to a lifetime, but as will become obvious, human emotional ontogeny mirrors human emotional evolution. My account, however, differs in that it construes emotion precursors as basic (in all three senses of Ortony and Turner 1990).

2.10 Summary

Emotions are still consensually considered to fall into two basic and distinct groups, despite recent challenges to this view. The two groups are basic emotions (BEs) and higher cognitive emotions (HCEs). BEs are characterized as genetically hardwired, as homologous to similar devices in other mammalian species, and as involving relatively stereotypical physiological, behavioral, and expressive patterns. In addition, they are also construed as relatively modularized and as activated by simple, unconditioned stimuli; they can also be activated by conditioned stimuli through learning. In contrast, HCEs are considered to be uniquely human; to lack unconditioned stimuli activation and stereotypical physiological, behavioral, and expressive features; and to vary culturally.

In both philosophy and psychology of emotion, theorists focus almost exclusively on full adult emotionality, construe emotions as operating primarily in individual psychological economies, and variously claim that emotions are or essentially involve cognitions, feelings, or action tendencies. In addition, they claim that they are either predominantly hardwired or socially constructed. Each view has merit, but none is adequate to explain the enormous complexity of human emotion. In chapter 4, I argue that emotions are essentially expressive-communicative devices that operate primarily in social economies, and both BEs and HCEs are language dependent. In addition, I argue that biology and culture interact continuously in human emotional ontogenesis.

3 Metaphysics and Mind

3.1 Introduction

In this chapter, using the lens of metaphysical realization, I focus on the contingent intracranialist–transcranialist debate. The term *contingent intracranialism* is due to Adams and Aizawa (2001, 48), who claim that it is a contingent empirical fact that cognition takes place exclusively in the head; Extended Mind theorists, hereafter termed *contingent transcranialists* (CTs; see chap. 1) disagree. As I noted in chapter 1, intracranialists accept that a great deal of cognition is heavily scaffolded by extracranial technological, natural, and sociocultural resources; it is the metaphysics of transcranialism with which they take issue. Indeed, they now claim that (i) the thesis of CT fallaciously conflates causal coupling with constitution, (ii) CT advocates fail to recognize the "mark of the cognitive," and (iii) these errors lead to methodological error. They also allege that the differences between internal and putative external processes are so marked that even transcranialists "maintain that cognition extends from brains into the extraorganismal world rather than from the extraorganismal world into brains" (Adams and Aizawa 2009, 92). I argue in this and subsequent chapters that the contingent intracranialists are mistaken on all four counts. I begin my rebuttal of the putative causal-constitution fallacy in this chapter and continue it in chapters 4 and 5.

As previously noted, my analytic device for this task is that of metaphysical realization, specifically, R. A. Wilson's (2004) views on context-sensitive realization. Given this, "context" from my three Cs (see chap. 1) features strongly. The context-sensitive view is the only one compatible with the theories of emotional and language development that I subsequently construct. I apply this device to two augmentations that transcranialists claim can extend cognitive systems, namely, technological and natural environmental resources, and I identify their component partial realizers, that is, core and noncore realizers in their total realizer systems. This allows me to demonstrate their bona fide mind-to-world transcranialism.

I include discussions of technological augmentations (e.g., a rake, which seems, prima facie, to be an unlikely candidate for inclusion in an extended cognitive system) and natural environmental augmentations (e.g., environmental objects), although they may appear strictly irrelevant to analyses concerning human emotionality. I include them for two reasons. The first is to establish that CT, in principle, extends over a range of different extrapersonal resources. Thus if CT, in general, is a flawed hypothesis, then CT regarding emotion, in particular, will be flawed. The second is because saccadic vision is highly influential, even indispensable, in language processing (see chap. 7), and this is highly germane to human emotionality. I argue that both of these augmentations are widely metaphysically realized; that is, they do involve a heterogeneous mix of intracranial and extracranial resources.

I contextualize my discussion of metaphysical realization by briefly reviewing the history of the individualist (intracranialist)–externalist (transcranialist) debate and contextualize the analyses included in the chapter as a whole by commencing with some introductory remarks concerning embodied, embedded, and extended cognition, all considered "species" of the "genus" situated cognition (Robbins and Aydede 2009, 3). I conclude the chapter with a comparative analysis of Rupert's (2004) hypothesis of extended cognition (HEC) and hypothesis of embedded cognition (HEMC). This provides a useful summary of the contemporary debate and a further opportunity to rebut some of the intracranialists' objections. I leave my rebuttal of the "mark of the cognitive" objection until chapter 6, where I marshal arguments against it.

3.2 Situated Cognition

According to situated cognitivist theory, mental activity is dependent on the situation or context in which it takes place, whether that situation be local (e.g., embodiment thesis) or more global (e.g., embedding and extension theses). These different "species" are more or less radical in both what structures and resources they construe as involved in cognition and how they are involved. Embodied cognition construes perceptual, cognitive, and motor capacities as dependent on features of the physical body, and embedded cognition construes cognition as heavily dependent on the natural and social environment insofar as a range of natural, technological, and sociocultural resources scaffold cognition (R. Wilson and Clark 2009). It is important to note, however, that both of these construals capture the explanatory importance of cognition's bodily and environmental dependence but do not commit to any constitution claims. Extended cognition, in contrast, is committed to constitution claims. Extended cognition, which includes vehicle or locational externalism, most recently termed contingent transcranialism (CT), claims, as the name implies, that genuinely cognitive processes and structures can extend beyond the skin and skull of the cognizing agent into her physical, sociocultural, and technological environments. These CT constitutivity claims depend, in turn, on claims that internal neural resources are deeply functionally integrated, to differing degrees, with these extraorganismic resources. In short, the theory of CT construes the supervenience base of genuinely cognitive processes and structures as sometimes extending into nonneural resources in the extraorganismal environment. According to this view, the metaphysical realization of such processes and structures is ineliminably context sensitive.

3.3 Embodied, Embedded, and Extended Cognition (CT)

As indicated earlier, these are differing species of the genus situated cognition, and as species, they differ in important respects.

3.3.1 Embodied and Embedded Cognition

Many forms of cognition are embodied in that they depend deeply on features of the individual's physical body; the body plays some significant causal or constitutive role in cognitive processes. These causal or constitutive processes take three forms, namely, to constrain, to distribute, or to regulate cognitive activity (R. Wilson and Foglia 2011). More germane to my focus in this book, however, is embedded cognition where "cognitive processes depend *very* heavily, in hitherto unexpected ways, on organismically external props and devices and on the structures of the external environment in which cognition takes place" (Rupert 2004, 393; his italics). Both intracranialists and transcranialists agree; where they differ is in their construal of the role of such external structures and devices. Intracranialists insist that such causally active structures and processes in the environment influence or act merely as input into intracranial cognitive processing (Adams and Aizawa 2001, 2008, 2009, 2010a, 2010b; Rupert 2004). These external structures and processes are never constitutive, that is, components of genuinely cognitive systems. What this implies is that transcranialists who argue that some cognitive activity is constituted by active features of the environment (e.g., Menary 2007, 2010) confuse the external causes of cognition with the external constituents of it. Although there are reasons to consider embodied and embedded cognition as leading to, or shading into, extended cognition, for example, in visual perception (A. Clark 1997), they are both entirely compatible with the denial of extended cognition. Both are compatible with neural and nonneural components being involved in inside-the-skin cognition.

3.3.2 Extended Cognition (Contingent Transcranialism [CT])

Extended cognition "occurs when internal and external resources become fluently tuned and integrated so as to enable the larger system—the biological agent and specific items of cognitive scaffolding—to engage in new forms of intelligent problem-solving" (R. Wilson and Clark 2009, 64). That is, CT "appropriates and integrates material resources around one into pre-existing cognitive structures. In cognition, agents modify or augment the capacities that those pre-existing structures enable" (58). Such modifications and augmentations differ in both their nature and their reliability and durability. In terms of their nature, they can be (i) natural or environmental, for example, where objects in the world are interacted with directly, as in visual perception; (ii) technological, for example, a rake; and (iii) sociocultural, for example, the close, linguistically mediated social relationship of neonate and caregiver. In terms of durability and reliability, they can range from (almost) permanent, for example, as in the development of new, or the elaboration of existing, neural circuits; or they can be temporary but repeated, fleetingly, time and time again, as in saccadic vision.

Arguments supporting CT have occurred in three waves (Cash 2013; Kirchhoff 2012). The first wave focused on functional parity; if the information encoded and retrieved from an external source, such as Otto's notebook (A. Clark and Chalmers 1998), was as easy to retrieve, as reliable, and as poised for automatic deployment as biologically encoded information, then it should be accorded similar status. Intracranialists disagree; what parity transcranialists construe as functional parity and CT, intracranialists construe as fallaciously conflating coupling with constitution. This fallacy, according to Adams and Aizawa (2009, 81), is to move from the causal coupling of some object or process to some cognitive agent to the conclusion that the object or process is part of the cognitive agent's processing: "What is common to these arguments is a tacit move from the observation that process X is in some way causally connected (coupled) to cognitive agent Y to the conclusion that X is part of the cognitive process Y."

Their claim, however, ignores the fact that no transcranialist has ever claimed that just any coupling of a cognitive agent to a noncognitive extra-organismic tool constitutes a cognitive system or agent, as the identification of Clark's parity criteria bears witness.

The second-wave arguments focus on the complementarity of the internal and external resources and their consequent integration into cognitive wholes (Menary 2007, 2010). Clearly what this implies, as Clark (2010) observes, is that CT depends on the degree of complementarity between internal and external resources and their degree of integration. If there is sufficient functional complementarity and integration, then it is plausible to treat the resultant larger system as a cognitive whole, with cognitive properties supervening on more than their biological components. Leading transcranialists (A. Clark 2008; Menary 2007; R. Wilson 2004) all claim that deep functional integration of nonneural resources in information processing is prerequisite to the formation of extended cognitive systems.

Third-wave arguments claim that both first and second waves imply the existence of distinct inner and outer realms, each with its own proprietary characteristics (see chap. 1). Third-wave theorists construe component structures, processes, and properties as continually changing in the process of cognitive assembly and agency. It will be clear from my analyses in the chapters that follow that my views are broadly consistent with those of the third wave.

3.4 Deep Functional Integration

It is important to note that the differing resources that transcranialists claim can augment cognitive systems do not function merely as inputs to such systems, as intracranialists insist. Rather, they augment, that is, alter the system by becoming deeply functionally integrated into what becomes an altered system. The distinction, it is claimed, is "intuitive enough, even if it is one that has been surprisingly difficult to pin down. … The *degree and complexity* of the integration is obviously implicated as is the goal or purpose of the system" in question (R. Wilson and Clark 2009, 63; my italics). They use the example of a radio, the purpose of which is to receive, decode, and play contents borne by radio signals. Anything that helps the radio to perform these functions better, such as a better transistor or signal amplifier, "looks like an augmentation of the system rather than mere input. This is so whether the additional transistor or amplifier falls inside the preexisting boundaries of the radio or lies beyond them."

This characterization, too, fails to pin down the distinction between augmentation and input, although it does get close. It also fails to answer the intracranialist challenge. Wilson and Clark (2009) are correct in suggesting that augmentations that improve function involve, because they necessarily must, some sort of alteration in structure and in processing. Radio augmentations in the form of new transistors and new signal amplifiers alter radio structure. The same is true of some cognitive augmentation: cognitive augmentations change neural structures in some way whether such augmentation is natural, sociocultural, or technological in nature. Intracranialists deny that this is augmentation, however; as noted previously, they readily concede that intracranial and extracranial resources can be deeply functionally integrated yet resist the further move to CT. They claim that such augmentations very heavily scaffold intracranial cognition but remain merely as causal inputs to it.

In my view, the move from DFI to CT does depend on the degree and complexity of functional complementarity and integration, as Wilson and Clark (2009) suggest, but this characterization is insufficiently precise. DFI and CT occur when the degree and complexity of functional complementarity and integration in extended cognitive systems are such that the internal and external components of the system function synchronically as a whole, producing effects resulting only from the system functioning synchronically as a whole. The more deeply functionally integrated extraorganismal resources are with intraorganismal ones, the more the system will function as a whole, and the more effects it will produce only by functioning as a whole. Inevitably, therefore, there will be cases of functional integration that are relatively stronger and relatively weaker in terms of complementarity, complexity, and integration. This will be particularly true when DFI results from a progressive developmental trajectory that depends on the development, maturation, and connection of neural substrata. In such cases, the functional integration of intraorganismal and extraorganismal components has to be triggered initially and then developed through processes in which functional complementarity, complexity, and integration are progressively increased. The process of emotional ontogenesis I describe in chapter 4 is instructive in this respect.

As noted earlier, the arguments of contingent intra- and transcranialists relate to the metaphysical realization of genuinely cognitive structures and processes. Before discussing forms that putative cognitive augmentations can take, therefore, I address the contemporary views and contextualize them in some brief introductory comments concerning individualism and externalism.

3.5 Individualism and Externalism: A Short, Potted History

One of the current orthodoxies in philosophy of emotion is that emotions occur at the interface of our internal concerns and the outer environment and, as such, are essentially what they are in virtue of their relational and functional properties. This orthodoxy, however, conceals what continues to be a deep division in philosophy of mind, that is, the division between individualists-intracranialists and externalists-transcranialists in terms of both metaphysical determination and how psychological states should be taxonomized and individuated. Very briefly, individualism arose as a means of "disciplining" psychology (R. Wilson 2004), that is, of establishing the conceptual, methodological, and institutional foundations for a science of psychology, distinct from philosophy and physiology from which it emerged (and from sociology, which was being "disciplined" at about the same time). Individualism was seen as a means of rendering psychology scientifically respectable. Individualism began as a normative constraint on psychological methodology, the principles of which were thought to be derived from the most successful explanatory practices in science. Such practices taxonomized their targets of investigation as supervening on the intrinsic physical properties of the individual object or organism that instantiated them.

This is "methodological solipsism" (Putnam 1975) and requires that psychological states be construed without reference to anything that falls outside the boundary of the individual who bears them. The motivation behind methodological solipsism was (and remains) twofold: it reflects the standard view (see hereafter) that (i) no psychological change can occur without some accompanying physical change in cognizing individuals; and (ii) how a cognizer conceives the world determines content, not how the world actually is (Bach 1982). For example, one can think of nonexistent things, of things and events long past, and of things and events yet to happen. According to methodological solipsism, approaches to cognitive science that are not individualistic are both methodologically and metaphysically misguided. They are methodologically misguided because the most explanatorily successful research paradigms in cognitive science and the sciences more generally have been individualistic, such as computational approaches to cognition and the taxonomizing of phenomena in science more generally according to their causal powers (which are a function of their intrinsic physical states). Contingent intracranialists, such as Adams and Aizawa (2001) and Rupert (2004), make precisely this claim. Nonindividualistic approaches are metaphysically suspect in that they must either (a) relinquish the most important contemporary insight into mental causation, that is, that cognition is computational, or (b) accept that the mind is not governed by principles that apply more generally to the physical world, such as supervenience on intrinsic physical properties (R. Wilson 2004, 13).

Individualists and externalists agree that agents causally interact with (a) their physical environments to collect information regarding it, using their different sensory apparatuses, and (b) their social environments, gathering information from their communicative interactions with others (R. Wilson 2004). The nature of their minds, therefore, and in particular the intentionality of their mental states, is at least partly causally determined by the character of their physical and social environments. In this account, the environment causally contributes to the contents of mind, in terms of what is perceived, thought about, and desired. Where they differ is that individualists insist that differences in the world cause some corresponding change in physical brain states, which give rise to differences in content, whereas externalists deny this. They insist that physically identical doppelgängers could have brain states with different content (and given that mental states with different content are ipso facto different types of states, their bearer's intrinsic physical states do not fix or determine her mental states).

The classic Putnam–Burge thought experiments of the 1970s sought to challenge individualism by demonstrating, respectively, the centrality of natural kinds and sociality to mental content. Taxonomic externalism arose as a result of these challenges; it continued to construe content-laden mental states as being internal to the individual who bore them, but allowed that they were not metaphysically determined (i.e., did not supervene on) the intrinsic physical states of the individual agent.

Putnam (1975) exploited the limitations of the descriptive theory of meaning of natural-kind terms to expose the limitations of individualism (R. Wilson 2004). Oscar, who lives on Earth, interacts with water on a daily basis; he drinks it, washes in it, and sees it fall from the clouds as rain. He has no conception of its molecular constitution; he describes it, therefore, as that which he drinks, washes in, and takes an umbrella to avoid. His doppelgänger, Twin Oscar, lives on Twin Earth. Twin Earth is exactly the same constitutionally as Earth except that there is no H_2O, only XYZ. XYZ is identical to water in appearance, taste, wetness, and so on. According to the descriptive theory of meaning, therefore, their terms *water* should have the same reference, but clearly they do not. Both Oscar and Twin Oscar stand in the same relation to their respective environments; that is, Oscar's water refers to water because it is drinkable, wet, and so on, and Twin Oscar's water refers to Twin Water because it has the same perceptible characteristics. What this means is that their natural-language terms differ in terms of reference despite the fact that their terms agree in their intracranial meaning. The reference of natural-language terms, therefore, is not fixed or metaphysically determined by what is in the head of cognizers; rather, it is fixed a posteriori by discovering the appropriate atomic constitution.

Burge's (1979) thought experiment does not rely on the semantics of natural-kind terms; it relies, rather, on construals regarding natural-language comprehension. Burge argues that we can and do possess incomplete understanding of the many things we think about and for which we have natural-language terms. Bert, for example, has incomplete understanding of arthritis; he thinks he has arthritis in his thigh muscle, that is, he thinks that arthritis is a disease of joints and muscles. In the actual linguistic community that Bert resides in, however, arthritis is a disease confined to joints, and this means that Bert harbors a false belief. In a counterfactual world, however, where arthritis does refer to a disease of both joints and muscles, Bert's belief would be true. This is interesting because Bert is physically identical in both worlds but has thoughts with differing contents. The linguistic community in which one is embedded, therefore, and not the intrinsic physical states of their bearer, metaphysically determines the content of one's mental states.

As previously indicated, the present debate more typically focuses on more radical species of externalism; these tend to subsume embodiment, enactivism, and distributed cognition (see above) and are variously classified as extended mind externalism (A. Clark and Chalmers 1998), vehicle externalism (Hurley 1998; M. Rowlands 2003, 2006), active externalism (A. Clark and Chalmers 1998), locational externalism (R. Wilson 2004), and environmental externalism (M. Rowlands 2003).[1] All of these explicitly recognize that the vehicles of mental content can include beyond-the-head relations and structures and, as noted earlier, are species of situated cognition.

Two important points are worth emphasizing. First, the theory of situated cognition is an ontological thesis, that is, a thesis of what mental processes are. (Of course, the thesis has epistemological implications insofar as understanding the nature of at least some mental processes depends on understanding an individual's manipulative, exploitative, and transformative operations on her environment.) Second, the deep functional integration of a range of nonneural resources into cognitive systems is always only partially constitutive of mental processes. There is always an ineliminable individualistic component, neural and bodily, in the constitution of mental processes (R. Wilson 2004; A. Clark 2008; M. Rowlands 2003).

3.6 Metaphysical Realization

The concept of realization (or, more accurately, realization physicalism; see Polger 2007) became a term of art in the 1970s through the work of Putnam (1975), for example, when it was introduced to formulate functionalism, that is, as a means of best characterizing mental roles. Early attempts used abstract computational specifications, later attempts used causal specifications, and still others used evolutionary specifications (Polger 2004). What these differing characterizations shared was the relational specification of roles in terms of input, interaction with other brain states, and outputs. For example, for my amygdala to realize the emotion of fear is for it to enter into relations with other brain states and the world through appropriate sensory-perceptual and motor mechanisms. Given the relational specification of roles, Polger (2007, 251) suggests that realization should be construed in terms of roles: "To occupy a role is to have relations that are distinctive of the role." In short, the realizer role of a property or state is to have the function of producing it:

R realizes G if R has the G function, if it plays the G role.

A group of neurons, for example, realizes the property of being an amygdala if it has the function of producing the emotion of fear.

The realization theory of mind began as the thesis that the brain should be understood as a computing machine, and the mind as a program or set of programs. This thesis reflects the view that computers implement or realize the programs or algorithms they run. Certain electrical states of the device realize computational states; the electrical activity of the device is not identical to any program state of Microsoft Word, but it implements or realizes such program states (Polger 2007, 236). In exactly the same way, it was claimed, mental properties are realized by physicochemical brain states but are not identical to them. Despite the importance of this claim, however, some scholars argue that realization has received little serious attention until recently (Polger 2004, 2007; R. Wilson 2001, 2004).

Realization is a relation of metaphysical determination (R. Wilson 2004, 2001); realizers determine the relevant realized properties. In Shapiro's (2004, 35–36) terms, a realizer R realizes kind K when K's existence depends on R and they exist synchronically. In other words, it is in virtue of R that K exists simultaneously with R. It is a "kind of compositional relation" (35) rather than a causal relation where the relation of R to K would typically be diachronic. Any adequate account of the realization of mental states must be able to account for the psychological, computational, and mechanical levels of realization. Today there are two views of realization physicalism (Polger 2007), namely, the standard view and the context-sensitive view. The standard view is espoused by intracranialists, and the context-sensitive view by transcranialists. I discuss these views in the following sections. The context-sensitive view, it is claimed, does more justice to the concept of realization operational in the biological, behavioral, and social sciences (R. Wilson 2004).

3.6.1 The Standard View of Realization

According to the standard view, realization is a two-place relation that holds between mental and physical states (R. Wilson 2001, 2004), where the physical states act as realizers for the mental. The intrinsic physical states of individuals, particularly of the central nervous system (CNS), exhaustively realize an individual's mental states. The standard view of metaphysical realization thus includes both a sufficiency and a constitutivity thesis:

(i) *Sufficiency thesis*: realizers are metaphysically sufficient for the properties or states they realize.
(ii) *Constitutivity thesis*: realizers of states and properties are exhaustively physically constituted by the intrinsic physical states of the individual whose states or properties they are (R. Wilson 2001, 4–5).

It is the conjunction of the sufficiency and constitutivity theses for at least a variety of properties and states, including mental ones, that Wilson (2001, 6) challenges:

> The sufficiency and constitutivity theses are not always true of the same putative realizers. Often the realizations that are metaphysically sufficient for the properties they realize are not exclusively physical constituents of individuals with those properties; conversely, sometimes the physical constitution of an individual with a given property is not metaphysically sufficient for that property to be present. Mental properties are no exception here.

The reason that the conjunction of sufficiency and constitutivity theses is often untrue of the same putative realizers, as previously indicated, is because realization is irreducibly context sensitive.

3.6.2 The Context-Sensitive View of Realization

Realizer systems of mental states are (obviously) metaphysically sufficient for the properties and states that they instantiate and can be either (i) entity bound, that is, confined within the boundary (skull and skin) of the cognizing individual, or (ii) widely realized, that is, extending beyond the boundary (skull and skin) of the cognizing individual, into her physical and social environments. In addition, both involve, in differentially extended configurations, nested contributory realizer components, termed *partial realizers*. The realizer systems that are metaphysically sufficient for the properties they instantiate are termed *total realizer systems*, and these include or contextually embed two types of partial realizers, which are termed *core realizers* and *noncore realizers* (to be defined below). Total realizer systems are also contextually embedded in a range of relevant background conditions, and this general contextual embeddedness points to the ineliminable role of context sensitivity in realization, both in general and in philosophy of mind, in particular (Hurley 1998; R. Wilson 2004). Take the adult experience of fear, for example: its core realizer is considered, by scholars of a somaticist-feeling persuasion, to be the amygdala because they view it as the specific part of the central nervous system most readily identified as playing a crucial causal role in producing or sustaining the experience of fear (R. Wilson 2004). They consider the amygdala to be the core realizer because they believe that its ablation or lesion eliminates or reduces emotional response. I argue in the sections that follow that they are correct in identifying the amygdala as core in adult emotionality but mistaken in terms of emotional ontogenesis. The process of emotional ontogenesis in human beings results in the development, maturation, and connection of the neural circuits, including the amygdala, which are the eventual bases of adult emotional experience. The core realizer in each case, that is, in emotional ontogenesis and adult emotional experience, differs.

This is important, for it points to the nature of core realizers; what is identified as core in any realizer system is a function of the conceptual or theoretical preferences of relevant scientists. Core realizers, therefore, are epistemically sensitive (R. Wilson 2004). Scholars of a distinctly cognitivist persuasion deny that the amygdala is the core realizer; they consider bodily and behavioral responses neither necessary nor sufficient for emotional experience. What is necessary for cognitivist philosophers and psychologists is a judgment or thought that something in the cognizer's environment (internal or external) is frightening (see chap. 2).

Two points concerning core realizers are worth emphasizing here. The first is that whatever core realizer is identified in any total realizer system, it is still only a partial realizer (Shoemaker 1984), because it must receive, at the least, some excitatory sensory-perceptual input. The amygdala, for instance, is activated by sensory input from thalamus and cortex, and without this input there would be no amygdala activation (Bush, Schafe, and Le Doux 2009). The second point is that, irrespective of conceptual disposition, core realizers are considered to provide the necessary conditions for emotionality. However, as partial realizers, they are not sufficient. When an individual adult is experiencing fear, as just indicated, other parts of her CNS (i.e., other partial realizers, such as the thalamus and cortex) are activated, and their activation is critical for amygdala stimulation to play the causal role that it does.

In sum, thus far: any property P is realized in a total system S that includes both core realizers and noncore realizers:

(a) Core realization of P: a state of the specific part of S that is most readily identifiable as playing a crucial role in producing or sustaining P.
(b) Total realization of P: a state of S, containing any given core realization and noncore realizations as proper parts, that is metaphysically sufficient for P (R. Wilson 2004, 108).

Rather more specifically: in the adult experience of fear, the total system, S, which realizes the property of fear, P, is the fear system, which contains, inter alia, receptors in different sensory-perceptual systems, particularly the retina, optic nerve, cortex, nuclei in the thalamus, amygdala, hypothalamus, medulla, and a range of neuroendocrine hormones. Fear (P) is a property experienced by some individual adult as a whole, but the total realizer (S) need not be identical to that adult, merely part of her. In terms of fear, for instance, the total realizer (S) excludes any properties of the individual who bears S that are not partial realizers of fear, such as her toenails or patellae. "Total realisations of P are exhaustively constituted by a core realisation of P plus … the non-core part of the total realisation" (R. Wilson 2004, 109). Noncore components of total realizers include the other parts of entities that are part of S; I have just noted that in the fear system, these include optic nerve, cortex, and so on. It is also important to acknowledge that other systems and processes that are not part of the total realization (S) are essential for the efficient functioning of S. These are a host of supporting background conditions that must obtain, for example, the provision of glucose and oxygen from digestive and respiratory systems, delivered to total realizer (S) by the circulatory system. Such background conditions are necessary for the efficient functioning of S and, indeed, for the efficient functioning of all body systems. Their input is necessary to develop and sustain total realizer systems such as S, but they are not part of S. Given their necessary input into the physical states constituting a total realization, however, it is, strictly speaking, total realizer systems plus background conditions that metaphysically suffice for P. In this account, even total realizations are metaphysically context sensitive.

Note that the maturation and connection of the fear system and the elicitation of fear, whatever else they are or involve, are both primarily mechanisms. Clearly, as the foregoing description of the components of the fear system demonstrates, they depend on a whole range of submechanisms (e.g., retinal cone stimulation, optic nerve depolarization, etc.) to operate as they should. In addition, as the description of the fear system's components also illustrates, they involve material entities (e.g., populations of neurons), but these crucially depend on certain spatial relations obtaining between them and their activation according to the correct temporal sequence (R. Wilson and Craver 2007). The fear system is composed of a hierarchy of mechanisms and submechanisms that are organized spatially and temporally to perform specific functions and subfunctions. I will argue, for example, in chapters 4 and 5, that submechanisms in the neonate are responsible for the unfocused crying and motor unrest (i.e., assistance-soliciting devices) associated with absolute threshold deficit, and submechanisms in the caregiver are responsible for the various behavioral components of her search-and-identify responses (i.e., assistance-providing devices). Further submechanisms in the caregiver serve to fine-tune her own assistance-providing devices, and these fine-tunings are fed back to the interacting neonate via further submechanisms to stimulate the maturation and connection of her specific assistance-soliciting devices and, eventually, her assistance-providing devices. If any of these processes fail to operate, their failure must needs influence the operation of other mechanisms in the hierarchy (as lesion and ablation bear witness).

As previously indicated, there is yet another distinction to note with respect to total realizer systems; the core and noncore realizers they include can be either entity bound or wide. The distinction between entity-bound and wide realizations may be summarized in terms of the location of the noncore component of a total realization:

(c) Entity-bound realization: a total realization of P whose noncore part is located entirely within the individual who has P.

(d) Wide realization: a total realization of P whose noncore part is not located entirely within the individual who has P (R. Wilson 2004, 112).

In entity-bound total realizations, for example, spermatogenesis, both core and noncore partial realizers (e.g., testes, testosterone) are, as are all total realizations, metaphysically context sensitive. Spermatogenesis depends on a host of necessary background conditions, for instance, a supply of oxygen, but spermatogenesis takes place entirely within the boundaries of sexually mature (or maturing) males. Wide realizations, which include partial realizers that lie outside the boundaries of their possessors, are also metaphysically context sensitive, and for the same reasons, that is, the necessity of certain background conditions obtaining. I argue in chapter 4 that human emotionality, including both emotional ontogenesis and adult emotional experience, is always widely realized; it is never simply entity bound, as intracranialists would assume. Wide realizations take two forms (R. Wilson 2004). In the first form, the total realizer system extends beyond the physical boundary of the subject that bears it; the core realizer, however, remains within the subject's physical boundary. In the second form, that of radically wide realization, the core realizer extends or lies beyond the physical boundary of the subject. Minimally, therefore:

(e) Radically wide realization: a total realization, the core realizer of which is not located entirely within the individual who possesses P (R. Wilson 2004, 116).

Wilson suggests that social actions like writing a check are paradigmatic of such realizations. In cases like these, the mere signing of a piece of paper constitutes signing a check because it occurs in a particular social system or practice, that is, a banking system; had the paper signing gone on in a different social practice (its total realizer system), it would have constituted a different action, such as signing a legal contract, marriage license, and so on. In addition, in cases like these, the core realizer, the actual signing of the check, clearly lies outside the boundary of the signing individual.

Two objections are possible at this juncture. The first relates to the threat of irrealism (R. Wilson 2004, 137), and the second to the threat of wide individuals.

(i) The threat of irrealism

Following Wilson, I claimed earlier that metaphysical sufficiency requires both that some physical system be in a certain state (i.e., a total realizer state) and that certain background conditions obtain, in short, that total realizer systems are metaphysically context sensitive. This view of total realizers underpins the realism that this thesis reflects. It could be called into question, however, if one could argue that total realizers, just like the core realizers they include, are also epistemically sensitive (because then mental states may simply be states ascribed to individuals). Indeed, one might even claim that, given that core realizers are epistemically sensitive, the total realizers of which they are partial realizers are epistemically sensitive, too. This claim, however, rests on the assumption that total realizers inherit their epistemic sensitivity from constituent core realizers, and this is not the case. A total realizer of P can be defined simply as a state of S metaphysically sufficient for P, ignoring any further claims concerning putative core realizers (R. Wilson 2004).

To be a realist about mental properties is to be a realist about their total realizers at least, and thus about the systems with respect to which total realizers are defined (R. Wilson 2004). The distinction between background conditions and noncore parts of the total realizer, therefore, is required for realism. Yet one might reasonably object that it is never possible to define precisely the boundaries of a total realizer system and its background conditions. Again, however, this is not the case. It is the case that the boundary between a total realizer system and its background conditions, in terms both of existence and of operation, can sometimes be unclear, but the individuation conditions for systems are highly constrained by the physical facts (R. Wilson 2004). For example, if the mammalian thorax were to be opened, the heart, great vessels, lungs, and pleura would be revealed, and scientists could choose to include the pleura, for instance, in the respiratory system and even the great vessels (at a stretch) or to view them as background conditions. It is how oxygen is taken into the body and carbon dioxide removed, however, that determines what, plus or minus extra components, constitutes the respiratory system. Anatomists discovered, they did not invent, what constitutes the respiratory system. Similarly, cognitive scientists discover, they do not invent, what constitutes realizer systems for cognitive processes, be they entity bound or wide.

(ii) The possibility of wide individuals

If mental states can be wide, why can't the individual bearers of such states themselves be wide? Quite simply because the characterization of realization I am espousing here preserves the view that it is an individual I that possesses P even though S may be wide or radically wide (R. Wilson 2004). Fear is a property of individual animals even though it is widely realized, and my thought that I should exercise more regularly is my thought. We attribute properties to individual persons and animals that we would hesitate to attribute to their component systems. Body systems have their own metaphysical constitution and have their own properties; they even contribute to the metaphysical makeup of their bearers, but they are not the bearers of some of the properties that their individual bearers bear. Digestive cells do not emote, plan, or grieve, but their individual bearers, in virtue of their other hierarchically and complexly nested properties, do.

3.7 Technological Cognitive Augmentation

I noted earlier that the deep functional integration into extended cognitive systems of an extraorganismal tool, a rake, appeared prima facie highly unlikely; some research evidence, however, relating to primate tool use and human unilateral visual neglect suggests otherwise.

Primate brains include bimodal neurons (Maravita and Iriki 2004, 79) that respond both to somatosensory information from a given body region (the somatosensory receptive field [sRF]) and to visual information from the space adjacent to it (the visual receptive field [vRF]). In a series of experiments, scientists recorded the bimodal neurons in the parietal cortex of Japanese macaques while the macaques learned to reach for food using a rake. After only five minutes of rake use, the experimenters discovered that some bimodal neurons whose original vRFs picked out stimuli near the hand had expanded to include the entire length of the tool, while others that previously responded to visual stimuli within the space reachable by the arm now had vRFs that covered the space accessible by the arm-rake combination. The researchers concluded that the vRFs responded to the rake as if it were part of the arm and forearm. Similar results have been found in human subjects suffering unilateral visual neglect where tool use (a stick) for reaching actually extends into the area of visual neglect to encompass the space now reachable using the stick. The researchers concluded that "simply holding the stick causes a remapping of far space to near space. In effect, the brain, at least for some purposes, treats the stick as though it were part of the body" (Berti and Frassinetti 2000, 415).

In this case, the total realizer system of the foraging system includes partial realizers of the rake and the macaque's body (minimally its hands, arms, shoulder, musculature, and nervous system) in terms of both neuromuscular and neurochemical supply and vRF. The core realizer in this case, that is, the part of the total realizer most readily identifiable as being centrally implicated in acquiring the food, is the macaque's hand-rake combination. Background conditions include all the circulatory, respiratory, and nutritional input into normal macaque physiology, plus the physical proximity (accessibility) of the food. Remarkably, however, by treating the rake as an extension of the macaque's natural arm, the macaque's brain treats this extended arm as the core realizer of food acquisition. Given this, the macaque's acquisition of the food is radically widely realized; the functional integration of internal neural and external technological resources could hardly be deeper.

In terms of locus of control, it has been suggested previously that the activities of macaques are controlled by enactive representations, and because of this, their locus of control, in terms of both triggering and execution, is shared between macaque and environment. I argue below that this is not the case, or not always the case for human beings.

Three points are worth noting at this juncture. The first is that such remappings and incorporations into body schemata might well explain the phenomenology of "tool expertise." Expert nurses with wound probes and blind persons with white sticks (Merleau-Ponty 1962), for example, typically claim to feel wound abscesses and environmental objects, respectively, at the end of their probes and sticks. This feeling would follow from the brain seemingly treating the probes and sticks as extensions of the agent's body. The second point is that such remappings and incorporations into body schemata tend to be only temporary, lasting only as long as tool-using, goal-directed activity persists; in terms of my analysis, however, this time frame is irrelevant. If neurons respond to extraorganismic tools as deeply functionally integrated into body schemata *at all*, as they clearly do, then they are deeply functionally integrated. Contingent intracranialists cannot object that this is just another example of very heavy Scaffolding 1, when the macaque's brain treats the rake as an extended component or constituent, that is, constitutive of the macaque's arm.

The last point relates to the comparative preadaptation of macaques and human beings to learn to fashion and, particularly, use artifacts in goal-directed activities. There can hardly be any comparison. Macaques are not preadapted to learn artifactual use to anywhere near the same extent as human beings. Their use of external physical objects is limited to the learned use of stones, sticks, and other natural objects found in their habitats. Human beings, in contrast, learn not only to use but also to fashion a range of artifacts that are not naturally found in their habitats. If human beings really are biologically prepared for artifactual-assisted culture, then we should expect the DFI of extraorganismal and neural resources, to differing degrees, including DFI with CT, to typify distinctly human activity, and I argue hereafter that it does. Any activity or interactivity that is partly realized transcranially exemplifies DFI with CT.

3.8 Natural Environmental Cognitive Augmentation

Research into visual perception provides compelling and remarkable evidence of this type of natural environmental cognitive augmentation; the evidence I refer to concerns research into change blindness and saccadic vision.

In change blindness, even extremely large changes in a visual scene, such as objects changing color or appearing and disappearing, go unnoticed, provided that the actual change is masked or the observer is not paying attention (Rensink, O'Regan, and Clark 2000; Simons and Levin 1997). In one experiment (Simons and Levin 1998)), the observer even failed to notice that the person with whom the observer was interacting face-to-face had been replaced by someone else. Keijzer and Schouten (2007, 116) note that this evidence explodes the subjectively plausible myth that seeing is like taking a photograph.

> A camera opens its shutter for a very short period, then closes it again, leaving a complete trace of the light pattern on its inner photographic film. Vision seems similar in that we (and our brains) seem to take in the whole visual scene with a single momentary glance, after which we have acquired a subjectively accessible internal representation of the external scene.

Change blindness suggests clearly that we do not construct rich internal representations of the external visual scene (for if we had, we would notice the changes). Visual perception is an ongoing process where the visual scene is scanned according to need, and at no particular moment is there a rich representation of the external visual scene activated in the brain.

This continuous scanning of the external visual scene is a function of saccadic processing. Only a very small area in the center of the retina (the fovea) is sensitive to visual details and to color, while the remainder of the retina generates a much coarser image. To compensate, normal vision involves four to five small eye movements per second, named saccades, which scan the visual field continuously. Only by and during such an active and ongoing process is the general detailed view of normal vision acquired. Vision seems intuitively an intracranial process, with the external world featuring merely as causal input to visual processing, but it is not. Rather, vision is physically constituted by the deep functional integration of intracranial and extracranial resources. It is only when these intra- and extracranial resources operate synchronously, as a whole, that is, when saccades scan environmental objects and events, that vision is realized. In this account, the core realizer of online vision is the interaction of neural and extraorganismic stimuli; online vision, therefore, is radically widely realized.

In terms of locus of control, vision in the human being can be controlled by symbolic representations, when the person is purposively scanning the external visual scene, or merely enactive, when she is not (i.e., when she is just scanning and monitoring the external visual scene routinely). The locus of control of her visual processing can thus change depending on the context in which it is operating (triggered by goal-directed motivation or routine monitoring), but the total realizer system and its core realizer do not. The individual human agent can symbolically trigger and represent the external world at will, but online visual processing is left to a heterogeneous range of internal and external resources. Online visual processing, therefore, is a transcranial achievement. What this implies is that if visual perception is involved in any online activity or interactivity, then the processes underlying them are partially metaphysically realized transcranially. This is true, as I will subsequently argue, of both language production and comprehension. It is also true, of course, of any emotional experience that is triggered or monitored in part by visual processing.

3.9 Sociocultural Cognitive Augmentation

Sociocultural extended cognitive systems are formed when an individual systematically and reliably depends on social structures, other people, and cultural products in her cognitive activities (Krueger 2013; Cash 2013).

They are so much part of mundane life that they are often taken for granted as being the normal conditions under which skills are acquired and facts learned. Indeed, R. Wilson and Clark (2009, 63) suggest that such extended systems are a kind of "cognitive oxygen." R. Wilson (2004), Clark (2008), and Donald (1991) all suggest that the most obvious cases are the writing systems that humans have invented and provide relatively durable public cognitive resources that are crucial to education, training, industry, and so on. Such systems are not just used by an agent, pace Adams and Aizawa and Rupert, but become functionally integrated into the cognitive functioning of the agent. They significantly augment the preexisting cognitive structures and processes to increase, for example, their long- and short-term memory capacity and focus on and critique their own ideas (A. Clark 1998). Mathematical notation, similarly, does not simply input into mathematical ability (although, of course, it does this); rather, it builds on such abilities to produce an agent with significantly greater mathematical ability (Menary 2013; Kirchhoff 2012; R. Wilson 2004).

There are many others, however, which again are so much part of the mundane furniture of the world that they are not recognized as extended cognitive systems. They are just accepted as part of the normal conditions under which their cognitive (including emotional) abilities develop. The various contingencies that are established in neonate-caregiver interaction in emotional ontogenesis and provide mind-world constancies that infants and children exploit (e.g., in social referencing, which I discuss in chap. 4) are useful examples here. Sociocultural resources

> are not biologically—genetically, physically, or evolutionary—givens, as they have been created and modified by the activities of past generations of people. So sociocultural resources are distinguished by their origin. But they are worth high-lighting in reflecting on extended cognition because ... they constitute a crucial part of some of the cognitive abilities and activities which distinguish human cognition from its nearest neighbors. There may be animal cultures, but it is only in *Homo sapiens* that we find diverse cultures of cognition, social structures and products that, whatever their own origins, now significantly augment the cognitive capacities of individuals embedded in them. (R. Wilson and Clark 2009, 63–64)

My own view, however, is that human emotional ontogenesis provides a more compelling example of the deep functional integration of sociocultural resources and intracranial circuitry and thus validates the CT hypothesis. This is because their concerted and synchronous operation, and only their concerted synchronous operation, results in emotional ontogenesis. I address emotional ontogenesis in considerable detail in the next two chapters.

3.10 Particular Intracranialist Challenges

In this section, I begin to address the alleged coupling-constitution fallacy and transcranialists' alleged methodological error. Further evidence that these intracranialist allegations are mistaken is provided in the next two chapters. I conclude the chapter with Rupert's comparison of HEC and HEMC; it provides a useful overview of some intracranialist objections and a further opportunity to rebut them.

(i) The Coupling-Constitution Fallacy

This fallacy, as Adams and Aizawa suggest, is to move from the causal coupling of some object or process to some cognitive agent to the conclusion that the object or process is part of the cognitive agent or part of the cognitive agent's cognitive processing. The *reductio* they offer to "terminally embarrass" transcranialists (A. Clark 2010, 81) is the following:

Question: Why did the pencil think that 2 + 2 = 4?
Clark's answer: Because it was coupled to the mathematician.

Their *reductio*, however, does Adams and Aizawa's cause a disservice; it shows with commendable clarity what is wrong with their objection. No transcranialist has ever claimed that just any coupling of a cognitive agent (mathematician) to a noncognitive extraorganismic tool (pencil) constitutes a cognitive agent (process or system). A mathematician who has been careless with a tube of superglue could end up being well and truly coupled to her pencil, but it would be implausible to claim that she and her pencil constituted a cognitive system or agent. Leading transcranialists (A. Clark 2008; R. Wilson 2004; Menary 2010) all claim that deep functional integration of nonneural resources in information processing is prerequisite to the formation of wide cognitive systems, and my previous analysis relating to macaque tool use demonstrates clearly that this is the case. Macaque tool use is an example of a genuine mind-to-world transcranialism. I return to the putative causal-constitution fallacy in chapters 4 and 5, to argue further that there is no such fallacy.

The now (in)famous example of cognitive extension, that of Otto, an Alzheimer's sufferer, plus his notebook (A. Clark and Chalmers 1998), may well lead to an inclination to overemphasize the coupling of cognizer and physical or technological tool in cognitive extension (as the foregoing *reductio* indicates); such an inclination, however, should be resisted. Extended cognitive systems include the deep functional integration of natural environmental resources, as previous analyses indicate, and, of particular importance for this monograph, sociocultural resources.

(ii) Transcranialists' Methodological Error

Adams and Aizawa (2001, 57) are concerned that transcranialism or extracranialism threatens scientific integrity. They suggest that combining intracranial structures and processes with extracranial structures and processes risks failing to identify the lawlike regularities that allow science to carve nature at its joints because these structures and processes are so different. They cite the physical rotation of the image in the Tetris game and Otto's notebook to claim that (i) firing electrons at a cathode-ray tube requires muscular activity to operate the button, and (ii) Otto's extended memory recall involves cognitive-motor processing not found in Inga's memory recall. More generally, we are invited to review the range of human memory augmenting technologies (photo albums, Rolodexes, notepads, etc.): "What are the chances of there being interesting regularities that cover humans interacting with all these sorts of things? Slim to none, we speculate" (Adams and Aizawa 2001, 61). In addition, in contrast, biological memory systems, as noted earlier, "display a number of what appear to be law-like regularities, including priming effects, chunking effects," and others. Transcranial processes are unlikely to give rise to interesting lawlike regularities. "There are no laws covering humans and their tool use over and above the laws of intracranial (inner) human cognition and the laws of physical tools" (61).

There are two points to note here. First, given that methodological solipsism explicitly excludes extracranial resources from the domain of the cognitive, and given also that it has been the dominant methodological doctrine for decades, *of course* there will not be many research results that include these. Second, there is a beginning research tradition that does recognize widely different resources in realizer systems. Clark (2008, 9), for instance, points to the successes of complexity theory in discovering unifying principles that apply across massive differences of scale, physical type, and temporality. He mentions the power laws that explain the emergent behavior of systems ranging from ant colonies to the World Wide Web. Analogously, he suggests that the processes that write to and read from Otto's notebook and those that write to and read from his biological memory could well be covered by a unifying framework of information storage, transformation, and retrieval.

3.11 Summary: The Hypothesis of Extended Cognition (HEC) versus the Hypothesis of Embedded Cognition (HEMC)

As noted earlier, Rupert's comparison of HEC and HEMC articulates some typical intracranialist objections to CT. Given this, it provides a useful concluding device for this discussion.

The hypothesis of extended cognition (HEC) claims that human cognitive processes and structures may loop into the environment in which the cognizer is embedded, and is thus to be contrasted with a more conservative view, the hypothesis of embedded cognition (HEMC), which is championed by Rupert (2004, 2010). Rupert claims that everything of scientific interest in cognition can be captured by such a view, and methodological error avoided.

Rupert presents a series of arguments in favor of HEMC:

(i) He appeals directly to common sense. He asserts that HEMC is more compatible with common sense than HEC; this implies that we should require strong reasons to prefer HEC.

(ii) However, such reasons are not forthcoming, because there are profound differences between inner and outer structures and processes (Adams and Aizawa make the same objection, but since it is the same objection, I did not mention it earlier); we should conceive of them, therefore, as different kinds.

(iii) Endorsing HEC would incur considerable scientific cost. HEC construes transient props and aids, available in the extraorganismic environment, as components (vehicles) of human cognitive processes. This risks robbing science of integrated, persisting, organismically grounded capacities, that is, those belonging to persisting biological individuals whose responses are tested in a variety of environments under a variety of conditions.

In addition, and of particular relevance to this discussion, even in terms of development, where change, not stability, is at issue, one still needs to find some persisting, developing core. HEC, it seems, "offers developmental psychologists no more reason to be interested in, for example, the series of temporal segments we normally associate with Sally from ages two to six than it offers to be interested in, say, Sally aged two, together with a ball she was bouncing on some particular day, Johnny, aged five, together with a book he was reading on some particular afternoon, and Terry, aged seven, plus the stimulus item he has just been shown by an experimenter" (Rupert 2010, 30).

What this obviously suggests is that the sciences of the mind cannot afford to identify human cognitive processing with the activity of various transient coupled systems comprising neural, bodily, and environmental elements. The cost would be the loss of progress that cognitive psychology has made to date.

Rupert's (2004) next objection relates to natural and explanatory kinds. This is concerned with the claim of Clark and Chalmers (1998) that Otto plus his notebook constitutes the supervenience base for some dispositional beliefs and therefore has more explanatory utility and power than Otto and notebook as Scaffolding 1. Rupert denies this; what constitutes a natural or explanatory kind is that which falls within the explanatory frameworks of successful science. Biological memory meets this requirement because it falls well within the explanatory frameworks of cognitive psychology, but extended memory does not fit the causal profile described by this successful science. Ergo extended memory is not memory at all.

My replies to these challenges must be obvious. First, we have no reason to expect that inner and outer structures and processes in any realizer system be similar; indeed, Clark and Chalmers's (1998) parity principle was formulated precisely to undermine the view that a fine-grained functional similarity between inner and outer processes is required. The claim made by Clark and Chalmers was that should an object in the external environment function in or as some process that, if it were to be intracranially bounded, would be accorded genuine cognitive status, then the fact that it is extracranial should be irrelevant.

> The parity probe was thus meant to act as a kind of veil of metabolic ignorance, inviting us to ask what our attitude would be if currently external means of storage and transformation were, contrary to the presumed facts, found in biology. Thus understood, parity is not about the outer performing just like the human-specific inner.

Rather, it is about equal opportunity: avoiding the rush to judgment based on spatial location alone. (Clark 2008, 114)

Second, we have no reason to expect that the various components in any unified realizer system should all behave according to the same laws; hormones, genetic products, salutatory conduction, and depolarization (to name a vanishingly small number of subprocesses and structures involved in emotional ontogenesis and elicitation) do not all behave according to the same laws, because they are different sorts of processes and structures (and, presumably, it is precisely these differences that account for the different contributions they make in different realizer systems).

Third, as Neisser (1981) reminds us, more than forty years of memory research undertaken within the frameworks of putatively successful science has harvested almost nothing of relevance to the understanding of everyday memory, that is, memory on the hoof. In memory research at least, the solipsistic methodology of traditional cognitive psychology, involving hypothesis testing in tightly controlled experimental conditions, has been neither explanatorily prolific nor powerful. Recognizing this sad failure (surely) has been influential in the move to extended memory.

Fourth, that some research programs are focusing on wide total realizer systems does not preclude others from focusing on partial realizer systems. The foregoing discussion about realization and function indicates that realizers are complexly and hierarchically nested from the intracellular level upward. Mutually informative research focusing on all these levels can (and already does) enable the expeditious and interactive growth of scientific knowledge.

Fifth, as indicated in chapter 2, there has been a progressive widening of conceptions of psychological development, from unidirectional, through bidirectional, to transactional or circular, and most recently to dynamical and coactional conceptions. All of these widen the focus of developmental psychology from the individual's neural circuitry to socioculturally sculpted developmental environments. I deal with these ideas in detail in chapters 4 and 5.

But what of Rupert's claim that we should preserve the persisting core of scientific focus, that is, the individual cognizer? Nothing in HEC threatens this core; indeed, as my earlier discussion of the possibility of wide individuals indicates, it remains individual agents who have thoughts, perceptions, dreams, and so on. In addition, and uniquely in the natural world, individual human cognizers are always the locus of control when voluntarily and actively engaged in information processing. This is true even when they are not the core realizers in radically wide total realizations (as when Sue signs her check). HEC, therefore, pace Rupert, just like HEMC, continues to construe individual cognizers as relatively stable spatiotemporally extended beings, albeit with porous boundaries.

In conclusion, we have no principled reason to eschew HEC; indeed, the opposite is the case. HEC offers cognitive scientists an exciting new way of conceptualizing the human cognitive agent, one that may prove more fruitful in generating more powerful explanatory frameworks.

4 Mirror, Mirror ... Human Emotional Ontogenesis

4.1 Introduction

In chapter 2 I reviewed, albeit briefly, the traditional views of human emotionality and three of the most recent contributions. I noted where my views differ from these more recent contributions and concluded, regarding the traditional views, that their emphasis on particular components of emotional experience resulted in each view failing to adequately explain certain emotional phenomena. I cited, for instance, the feeling theories as eschewing the importance of cognitions and cultural variation, and cognitive theories as eschewing the importance of subjective feelings and physiological reactions. I suggested that these different construals reflect a more fundamental construal concerning the nature of emotions, namely, whether they are primarily biology-nature or culture-nurture. I claimed that the developmental account that I have constructed addresses all such limitations and also provides some insight into the mechanisms underpinning emotional development. This is the first of two chapters that present my account and, as noted in chapter 1, begins to reveal the depth of the complementarity and integration of intracranial and extracranial resources involved in emotional ontogenesis.

In chapter 3, I began my rebuttal of two intracranialist challenges to transcranialism, namely, that by causally coupling intra- and extraorganismal resources, transcranialists fallaciously conflate causal coupling with constitution, and the differences in these putative resources are such that cognitive extension is always from brain to world. I provided two compelling examples of mind-to-world DFI with CT, one relating to technological augmentation, and the second relating to natural environmental augmentation. In this chapter I provide a further, compelling example of world-to-mind DFI with CT relating to a sociocultural cognitive augmentation, that of human emotional ontogenesis. I provide further evidence in chapter 5 when I deal with language acquisition. I describe the processes involved in such rich detail that any intracranialist denial should be obviated. An incredulous stare, as David Lewis was apparently fond of reminding us, is no substitute for argument (Millikan 2004).

The example of DFI with CT that I describe concerns the concerted operation of intracranial and extraorganismal sociocultural resources to produce new or reshape existing neural substrata, including interneural connection, and full human emotionality. Drawing on insights from developmental psychology, I argue that there is no illegitimate conflation of causal coupling and constitution in CT accounts of human emotional ontogenesis. The synchronized and mutual modulation of the relevant causal processes in caregiver and neonate, to which both are preadapted, provides the necessary and sufficient conditions for the development of full human emotionality, but only within a close, linguistically mediated social relationship. In addition, I argue that the mutual modulation of the differing causal processes in caregiver and neonate results in the development of the neural substrata of discrete emotions in neonates and children and the refinement and elaboration of the neural substrata underpinning caregivers' intuitive parenting skills. In light of these analyses, I also argue that human emotional ontogenesis and experience are crucially dependent on the interplay between biological equipment and sociocultural environment. In addition, I argue that it is a mistake to construe the directionality that transcranialism takes as always from brain to world; the ontogenesis of human emotionality is a world-to-brain transcranial achievement.

My analyses constitute an argument that human emotions develop to enable human social life, first through interpersonal regulation, and subsequently through intrapersonal regulation. Human neonates and their primary caregivers are genetically and biologically equipped with a limited but perfectly complementary range of sensorimotor constraints and species-typical behavior patterns that enable the development of the very close, linguistically mediated social relationship that, in turn, provides the context for the development of increasingly complex forms of behavior. This relationship provides the necessary sensory-perceptual stimulation to excite genetic activity and the resultant development and maturation of emotion-related neural substrata. I describe in detail the nature of the sensorimotor constraints primarily involved in emotional ontogenesis; in chapter 5, I describe in similar detail the species-typical behavior patterns (STBPs), the interaction of which contributes to the development of human emotionality. The interaction of STBPs also provides the basis for the development of language and symbolic thought, which are essential to the development of full human emotionality. I conclude the chapter by assessing intracranialists' putative acceptance of some forms of DFI.

4.2 The Ontogenesis of the Emotions

"Organisms inherit environments along with their genes. ... Parents create very specific conditions for infants and the adult is the product, not of its genes, but of the interaction of its genes in this structured developmental context" (Griffiths 1997, 61). This is particularly true in relation to mental capacities. Ample evidence accumulated over the last forty years demonstrates that both human and nonhuman primates will fail to develop normal adult mental capacities and certain normal adult behaviors if denied the social interaction to which they are preadapted (Harlow and Harlow 1962; Harlow and Zimmerman 1959). Emotions evolved to control and regulate the behavior of members of social species (Vandekerckhove et al. 2008), and it is a necessary condition of emotional ontogenesis that social animals are exposed to the social lives their emotions evolved and developed to regulate.

The environments into which human neonates are born vary enormously across the globe; it is entirely reasonable, therefore, that they should have to learn which objects and events in them should elicit emotional responses. It is unremarkable that although emotion-eliciting stimuli are learned, their learning is preadapted by the provision of precursor emotions that evolved to deal with experiential saliences in ancestral environments. Such saliences clearly included stimuli that were dangerous, noxious, challenging, novel, and pleasurable. These match the emotion precursors (affect expressions) with which human neonates are genetically equipped, that is, fright (danger), disgust (noxious stimuli), distress (challenges), interest (novelty), and endogenous pleasure (pleasant sensations). These affect expressions serve to benefit the neonate directly in its relationship to the environment; for instance, disgust results in opening the mouth, wrinkling the nose, extruding the tongue, and retching, which removes noxious substances from the digestive tract. In human emotional ontogenesis, however, they also benefit the neonate indirectly by serving as signs of physiological and psychological status to caregivers, for example, noticing signs of disgust during feeding and, as a result, interrupting feeding. This response presupposes correctly interpreting the sign and then providing the appropriate intervention. It is enabled by the complementary preadaptation of caregivers, which takes the form of intuitive parenting skills.

As a result of this complementary preadaptation, human emotional ontogenesis can be construed as taking place in five phases, three of which are germane to this book. These phases are age-related:

- Birth to two years: During their first two years, neonates and infants build a differentiated and culturally specific repertoire of emotions from the affect expressions with which they are monomorphically and universally endowed. They learn to differentiate basic emotions as expression signs and acquire a repertoire of appropriate coping actions, all within the framework of interpersonal regulation with caregivers.

This implies that, at birth, neonates have components of emotional systems (appraisal, body reactions, expressive reactions, and feelings) only in precursor forms; this further implies that maternal causal attribution processes and coping actions are precursor, too.

- Three to six years: During this period, young children progressively reduce the comprehensive support previously required from caregivers and become capable, gradually, of intrapersonal regulation (Sroufe 1996). The emergence of the self-evaluative emotions, such as pride, shame, guilt, and embarrassment, connect the actions of the increasingly self-aware child to her culture's norms and values. In addition, children show less inclination to have their goals satisfied by others and are able, increasingly, to delay gratification. In short, they learn to accommodate their motive satisfaction within cultural norms and values and to prioritize motives and goals and delay or even suppress gratification.
- Six years to adolescence: Expression and speech signs are used increasingly for intrapersonal regulation.

The ontogenesis of human emotions demonstrates that they develop from emotional precursors (affect expressions), but only in the context of the appropriate human social relationships. Moreover, both human neonates and their primary caregivers (usually mothers) are genetically preadapted; that is, they coevolved to enable such appropriate human social relationships. Human infants will develop into emotionally blunted children and adults if denied the appropriate socioemotional developmental context.

Human neonates are the most neurologically immature of mammalian neonates (Prechl 1993), and this is particularly true in relation to their motor abilities (Thelen 1984). In light of this, and to ensure survival and well-being, certain competencies evolved via a process of natural selection in both neonates and caregivers. In neonates, these take the form of a limited range of affect expressions, precursor strategies for regulating emotions, and sensorimotor competencies; and in their caregivers, they take the form of "intuitive parenting" skills (Papousek and Papousek 1987).

Affect expressions, that is, the referents of the English words *distress, disgust, fright, interest,* and *endogenous pleasure* (Izard 1978), which are consistently observed in neonates from birth, are triggered solely by absolute stimulus thresholds; they are not directed toward particular objects. They are expressive reactions to sensed positive or negative internal physiological change (e.g., hunger, pain, overstimulation) or external stimuli (e.g., body contact, sensory overarousal) that presuppose a further innate endowment, that is, interoceptive and proprioceptive receptors in viscera, muscle, and skin, which are crucially important in emotional ontogenesis. The function of distress, disgust, and fright is to signal needs-related deficits (e.g., hunger) or impairments to physical integrity to caregivers (Sroufe 1996). The function of interest and endogenous pleasure, however, is to help to build representations of the external and internal physical and phenomenological environment. In terms of precursor strategies for regulating emotion, neonates are equipped with only two, and these operate only within a limited arousal range. Neonates engage in sucking to calm themselves and look away from overwhelming stimuli (Blass and Ciaramitaro 1994) to control arousal. When arousal exceeds its stimulus range, the infant shows signs of (i.e., expresses) distress. Distress is typically elicited by a threshold-exceeding deficit state and is expressed initially through motor unrest. This is followed by unfocused crying, which increases in volume if distress continues or increases. Neither the motor unrest nor the crying indicates which threshold-exceeding deficit state is causing the distress

(Lester 1984). Caregivers have to discover for themselves what ails their infant and what to do to remediate it.

What these distress expressions do, depending on their relative sensitivity, is elicit in caregivers a measureable degree of psychophysiological arousal, that is, measurable physiological stress symptoms (Boukydis and Burgess 1982), feelings of intense pressure to give assistance until the crying ceases (Gustafson and Harris 1990) and to seek the cause of distress, remove it, and engage in intuitive calming actions (M. Papousek 1990). Sensitivity means that caregivers perceive their infants' precursor emotions, interpret them correctly, and react to them appropriately; it is therefore the basis of successful intuitive parenting. Sensitive caregivers perceive unspecific and subtle cues from their infants and respond promptly, regardless of their own current motives. Sensitive parenting results, at three, six, and nine months, in reduced negative affect expression, reduced frequency of crying, and increased communicative abilities (Ainsworth and Bell 1974).

In rather more detail: according to Papousek and Papousek (1999), intuitive parenting typically takes three forms. The first is that of coregulation of neonatal "emotion." If the infant is distressed, sensitive parents intuitively engage in appropriate calming activities, and when the infant is apparently pleased with something, the parent attempts to prolong the stimulation and also to intensify it. The calming activities in which parents engage are precisely those to which neonates are preadapted to respond, that is, seeing a human face, hearing a human voice, being cradled in human arms. These activities are aimed at optimizing the infant's state of arousal to optimize interaction opportunities. Sensitive caregivers register the state of the infant's arousal and adjust their behaviors accordingly. The caregiver assesses the baby's arousal level by opening her mouth or her hand to assess muscle tone; a lax tone indicates fatigue, but a good tone indicates readiness for interaction.

The second form that intuitive parenting takes is the direction of infant behavior in optimizing the opportunity for motor mimicry. Parents express themselves with succinct, exaggerated expressive reactions, for example, exaggerating a facial expression and modulating their voices carefully in "baby talk" ("motherese"). These exaggerated expression signs exaggerate the central features of human emotional expression and the pragmatics of human speech (see chap. 5). The third form that sensitive parenting can take is allowing the infant's expression signs to direct their behavior. When parents intervene to provide whatever it is they believe their infant wants or needs, the baby responds with positive or negative feedback. Positive feedback includes smiling, cooing, and snuggling up to caregivers, and these responses encourage parents to provide the same response in similar situations, which, in turn, enables more prompt and more sensitive intuitive parenting. Thus the infant's expressive reactions and caregivers' intuitive parenting constitute a positive feedback loop that results in similar assistance provision in similar assistance-soliciting situations. This behavior results in the progressive fine-tuning of parents' assistance-providing devices.

It is important to note, however, that sensitive intuitive parenting is not simply reactive. It is also proactive in selecting appropriate environments to optimize infant arousal and in actively seeking contact with the infant from birth. Caregivers talk, smile, cuddle, and position themselves to make mutual gaze possible (which engages precisely those resources with which infants are innately equipped). What all of this implies is that infants are very poorly equipped for intrapersonal emotional regulation but very well equipped for interpersonal regulation with caregivers who are preadapted to interpret their needs and respond appropriately.

Interest and focused attention are triggered by novelty of external stimulation, and novelty is tied to certain temporal, sensory, and spatial contingencies. Neonatal visual perception is limited to 20 to 25 centimeters and exhibits a preference for face-shaped forms (Umilta, Simion, and Valenza 1996), and neonatal auditory perception exhibits a preference for the frequency, range, and pattern of human speech (M. Papousek 1994). The "speaking" and slowly moving face of the person holding the baby in her or his arms (Langsdorf et al. 1993), moving slowly enough to track (Brazelton 1983), is therefore one such contingency. Neonates do not simply respond to such contingencies, however; they also actively search for them (Stern 1985). When interest is stimulated, the baby ceases unfocused motor activity, orients toward, and visually fixates the stimulus object (Malatesta and Wilson 1988), which signals receptivity for information and interaction to the caregiver. The caregiver assesses the arousal state of the baby and her readiness to interact and adjusts her facial, vocal, and gestural behaviors accordingly.

I stated earlier that human neonates are born equipped with certain sensorimotor competencies, and cited sucking and turning the head as two of them. A third, critically important competency is that of motor mimicry (Meltzoff and Moore 1988, 1989).[1] Neonates imitate the facial expressions of caregivers, including sticking out their tongues and opening their mouths and even facial expressions for emotions, such as smiling,[2] frowning, and pursing their lips, within two days of birth. What this seems to imply is that neonates can translate visually perceived motor patterns into the appropriate proprioceptive ones. The mimicry competence of neonates corresponds to a complementary competence in caregivers. They adapt their communication patterns intuitively to the infant's perceptual limitations to reduce complexity; they mirror babies' expressive reactions in their own expression and use "exaggeratedly succinct" (Holodynski and Friedlmeier 2006, 100) expression signs to enable the infant to experience contingencies between expression and feeling (Gergely and Watson 1999). The expression is so exaggerated, for instance, an exaggerated smile, that it elicits an imitated smile (with its accompanying feeling) in the infant.

As previously noted, in the first two years of life, infants must master two tasks through interpersonal regulation with caregivers: (i) they must construct, from the affect expressions and sensory-perceptual constraints with which they are preadapted, a range of differentiated emotion systems mediated by expressive reactions such as joy, anger, and sadness; and (ii) they must acquire a repertoire of appropriate coping mechanisms. The emotion systems are composed of appraisal, expression, body reaction, and feelings, which means that the framework of interpersonal regulation must enable both the synchronous differentiation of these components and their appropriate interaction. It is the componential differentiation and interaction that leads to the development of differentiated sign-mediated emotion systems in infants. The development of these systems allows infants to assume, progressively, an increasingly equal share of interpersonal regulation by enabling their caregivers to satisfy their changing needs promptly and precisely.

It is through sensitive and prompt nurturing that carers create contingencies between emotion elicitors, their infant's precursor emotional responses to these, their own interpretation of the infant's needs, and their own actions to deal with them. These contingencies constitute mind-world constancies that are exploited by the neonate to fine-tune her producer devices in emotional ontogenesis and by the caregiver to fine-tune her interpreter devices. Through this process, the differing components become integrated into differentiated, sign-mediated emotion systems in neonates. In addition, the process allows infants to seize control of interpersonal encounters by orienting and guiding caregivers precisely and promptly to fulfill the infants' needs. Infant smiling provides a good example. Caregivers are preadapted to seek contact with their baby and to position themselves to make mutual gaze possible; they also talk and smile in an attempt to elicit a reaction. The mother smiles, and the baby imitates it, to the delight of the mother. Parents respond by increasing their own smiling and vocalization and thus establish the conditions within which the infant will build up contingencies between smiling, its eliciting causes, and the feelings of pleasure that become associated with it. That smiling is an expression of pleasure develops through synchronous experience of appraisal (cause) and feeling.

The developing ability to differentiate emotion elicitors is evident in the changes in infant smiling behavior in the first three months (Sroufe 1996, 81). Smiling in the neonate is evident some seven seconds after perception of the cause and manifests in lifting the corners of the mouth, sometimes with the eyes closed. By three months, however, the infant smiles promptly after stimulus perception, directs her attention to the stimulus object, and actively grins and vocalizes or coos (81). What this reflects is the countless opportunities offered in the first three months for face-to-face interactions in which caregivers smile at infants to encourage motor mimicry and the learning of relevant contingencies.

Similarly, emotional reactions differentiate progressively in the early months, and anger provides a useful example. In their study of one-, four-, and seven-month-old infants whose arms had been restrained to induce a negative emotional reaction, Sternberg and Campos (1990) found the following:

- At one month, infants reacted with a series of undifferentiated negative facial expressions and unfocused gaze before starting to cry.
- At four months, they reacted with a clear expression of anger: drawn eyebrows, open rectangular mouth, and sometimes narrowed eyes; and at the commencement of the restraint, their gaze was focused on the hand or face of the person restraining them. This reaction demonstrates that they had identified the cause of the discomfort.

- At seven months, infants directed their gaze to their attendant caregivers, demonstrating that their anger had acquired a socially directed appeal character.

What these findings strongly suggest is that unfocused affect expressions of the neonate are transformed into specific basic emotions that are directed toward their cause. These follow the cause promptly, are fine-tuned to their context in terms of focus and timing, are supported by coordinated body reactions, and trigger assistance from caregivers.

It is worth noting, however, that not just any combination of appraisal pattern and expression sign can be generated within the framework of coregulation. Expression signs are typically coded iconically in line with the action readiness triggered by the appraisal pattern. Anger, for instance, requires the mobilization of physical resources needed to engage in aggressive confrontation. In infants, these include scratching and biting opponents. Instrumental acts of aggression can then lead to the emergence of iconic signals of readiness to attack in the hope that the opponent might back down. Raising muscle tone, baring teeth, and clenching fists then become iconic expression signs for anger and are likely to be interpreted as such. This implies that the forms in which expression signs are encoded as icons and the instrumental usefulness of many body reactions impose limits on how expression signs and appraisal patterns can be combined. This might explain why some facial expressions are interpreted and used in the same way in different cultures. This in no way implies, however, that such expression signs have not been fine-tuned in interpersonal regulation (Holodynski and Friedlmeier 2006, 107).

It is the interaction of the caregiver's affect mirroring of infant expression reactions and infant motor mimicry that leads to the development of context-coordinated, emotion-specific appraisal patterns. The infant's rudimentary expression imitation is taken up by sensitive caregivers and shaped into salience through caregivers' succinctly exaggerated affect mirroring.

Motor mimicry of emotion-specific expression signs is observed in infants as young as 2.5 months (Field et al. 1982). Infants show increased happiness and anger expression signs in response to maternal happiness and anger expression signs. Remarkably, however, when exposed to maternal sadness expression signs (drooping mouth, raised eyebrows), infants respond with clear sucking and mouthing movements. This seems to suggest that infants of this age are not merely mimicking expression signs but beginning to feel the emotion (sadness). In addition, infants as young as two to three months recognize the contingent affect mirroring of caregivers and, in response, even anticipate such contingent affect mirroring (Legerstee and Varghese 2001), and this leads to a synchronization of caregiver and infant expression signs.

It is worth emphasizing that the reciprocal influencing of appraisal patterns and expressive reactions is generated by the special context in which infant emotions develop: Fogel (1993) terms this interdependent relationship *coregulation* to emphasize the interdependence. Infants build up their emotion-specific appraisal patterns not just in physical space but also in semantic space in which their emotional experiences are mediated by the interpretations of caregivers. Infant expressive behaviors play an important role in the emergence of these interpretations, not just as instrumental adaptation reactions to the physical environment, but also as indexical and iconic signs that appeal to the caregiver. Caregivers infer their infant's emotions and intentions through the coincidence of contextual features, knowledge of their infant's developing character, and her current expressive and body reactions, and they coordinate their responses with this interpretation of their infant's expression. In turn, caregivers infer the rightness or wrongness of their interventions based on their interpretation of feedback from the infant's expressive and body reactions.

As noted in chapter 2, this interdependence of neonate and caregiver in human emotional ontogenesis has increasingly been acknowledged in developmental psychology since the early 1990s, for example, Fogel's (1993) coregulation model, and, if Vygotsky's work is included, since the 1930s. Conceptions have progressed from unidirectional models (where parental behaviors were viewed as antecedents and neonatal behaviors as outcomes), to bidirectional models, to transactional or circular models, and most recently to coaction models and explicitly dynamical models. The earlier models have progressively been challenged as relatively impoverished in explanatory power, especially since the development of sophisticated microanalytic videotaping and computer technologies. Such technologies capture the synchronized, continuous parent-child-parent relationship in which it is impossible to pinpoint the influence of one partner on another at any given moment. This is demonstrated in mutual gaze interactions: the infant's gaze elicits the caregiver's gaze, which perpetuates the infant's gaze, and so on; in short, the two gazes are both effects and causes simultaneously in mutual gazing, because they each affect the other's nervous system simultaneously. As Schore (1997, 602) puts it, this is evidence of "mutual entrainment of mother's and infant's brains." It is worth emphasizing that coaction denies that it is the additive effects of different developmental influences that cause development. It is, rather, the *relationship* or coaction between two or more components, not the components themselves (Gottlieb and Halpern 2008; their italics).

Clearly, therefore, to insist that parental behaviors, including their linguistic behaviors serve as mere input or triggers to a cascade of developmental processes, as intracranialists most certainly would, is to fly in the face of state-of-art developmental psychology. Parental behaviors are no longer considered as mere antecedent causes of infant outcomes or effects because there is burgeoning research evidence of synchronized coaction between caregivers and neonate.

A further transformation accompanies the emotion-specific differentiation of expression signs: their emotion-specific conventionalization during interpersonal regulation transforms expressive reactions into expression signs. As conventionalized signs, they represent generalized emotion-specific responses. An expression sign does not just stand for a simultaneously triggered emotion but can also, just like a word, be used to represent the emotion symbolically. It can then be used purposefully, deliberately for affect attunement, social referencing, play, or even deceiving an interaction partner.

One of the clearest examples of conventionalization of expression and its symbolic use is in the social referencing manifested by infants from about ten months (Walden and Baxter 1989). An infant seeking assurance from her mother perceives her facial expression not as a mirror of the infant's own emotional response but as a symbol for one. For example: A stranger offers a toddler a cookie in her mother's presence. The toddler looks to the mother and reads from her facial expression how she should react. If the mother smiles, the toddler accepts the cookie; if, however, the mother looks anxious, the toddler refuses the cookie.

This constitutes an enormous achievement for a ten-month-old (Holodynski and Friedlmeier 2006, 111) and reflects a great deal of learning since birth: the toddler has already learned that (i) she can attract her mother's attention simply with a questioning gaze; (ii) this attention can be focused on feelings, in this case, the toddler's own uncertainty about what to do; (iii) facial expressions provide information about feeling (which she learned through affect mirroring); and (iv) her mother's expression in this situation reflects neither her own nor her toddler's feelings. It provides an answer to an unspoken question concerning what to do. Smiling means "everything is fine—go ahead," whereas an anxious expression means "careful—don't do it." All of this, remarkably, is managed without speaking at all, just by using expression signs symbolically.

By ten months, therefore, expression signs can be used to indicate what is the case and what one should do. The mother's smile is taken not as an expression of her real feeling state but as a symbol that has become detached from it. Nevertheless, this first symbolic use of expression signs is still tied completely to the situational context and, initially, to approach or avoidance. The expression sign has not yet acquired the multifaceted, generalized meaning that it possesses for adults.

The symbolic use of expression signs is also found in mimic emblems (Ekman and Friesen 1969) and in symbolic play. Mimic emblems are conventionalized mimic signs that are ascribed specific meaning with which they regulate social interaction; raised eyebrows to indicate a question to a partner is a common one. In symbolic play, infants imitate prototypical expressions, such as imitating a baby's whining or crying or the comforting behaviors of parents in infant-child interaction. Another exquisitely mundane example, well known to most parents, is the symbolic use of distressed howling when a wish is denied, to be followed immediately by its cessation once the wish has been granted. This howling is not a symptom of real distress but merely a symbol of it, which is used strategically and instrumentally by young children (especially when howling is effective in producing the desired result).

It is important to note that before infants can use expression signs symbolically, they have to perceive and acquire an internal representation of contingencies between the cause of an expression, the expression itself, and what to do about it. (In addition, a further set of contingencies must be built up so that an expression sign can be detached from the concrete emotion episode and used as a symbol, that is, to represent the emotion.) There are contingencies between the sensations in infants generated by their own expression signs and the expression signs of caregivers that mirror the infant expression signs in succinct, conventionalized ways. The mirrored expression thus becomes a sign marking the currently experienced feeling state that has acquired an emotion-specific meaning for the infant through the repeatedly experienced contingency between cause, feeling, and action. The smile that the caregiver uses to mirror the infant's smile refers to the infant's proprioceptively sensed smile. For the infant, this sensed smile is connected to further typical sensations for the emotion of joy or pleasure such as relaxation. It is also linked to a cause such as seeing the caregiver again and to an action to deal with it, namely, to enable the state to persist or to initiate it again.

Human neonates are born with just two emotion-regulating strategies, these being averting their gaze and sucking, but because these are successful in only a limited arousal range, emotional regulation develops predominantly in the form of interpersonal regulation between infant and caregivers. Parents, as noted, strive to maintain their infant's optimal arousal levels, stimulating them when they show lack of interest, and calming or distracting them when they are distressed. In these responses, parents tend to use the same strategies that are modeled by their infant, for instance, a pacifier and mobile or playful distraction. In addition, parents resort to antecedent emotional regulation by preselecting contexts that are more likely to be optimal arousal contexts (R. Thompson 1990).

Of course, in the first two years, toddlers are characterized by the need for instant need gratification; they are tied exclusively to the here and now. Nevertheless, they do use strategies that exercise a degree of intrapersonal control; these include calming (sucking, comfort blanket), distraction (moving away, object manipulation), and, rarely, symbolically mediated strategies (reinterpretations) and antecedent strategies (purposeful selection and control of contexts that provide positive and avoid negative emotion). All of these are first applied during interpersonal regulation.

Affect mirroring elicits a further major transformation involving the feeling component of an emotion: the emergence of conscious feelings. Through affect mirroring, the unfocused sensation of interoceptive and proprioceptive feedback in the neonate becomes the conscious feeling in the infant. The all-encompassing, unfocused sensation of body and expressive reactions induced by an elicitor is transformed into emotion-specific organized feeling. Characteristic of the latter is that from the range of simultaneous interoceptive and proprioceptive sensations, it singles out and emphasizes the sensations that are typical for the specific emotion and symbolizes them through an expression sign.

For the feeling state of pleasure, the expression sign that is also felt proprioceptively is smiling. At the same time, further feedback sensations are felt, such as relaxation and warmth, vocal gurgling, or effusive movements that all characterize the subjective feeling of pleasure. In contrast, other occurrent feedback sensations, such as a painful diaper rash, are not classified as pleasure feelings. "Just as the formation of symbols in speech signs categorises and structures perception of the external world, the formation of symbols for expression signs links together and structures the perception of the internal world, that of interoceptive and proprioceptive sensations" (Holodynski and Friedlmeier 2006, 113). Gergely and Watson (1999) suggest that this transformation process is analogous to that of adult biofeedback therapy (Holodynski and Friedlmeier 2006, 113). Biofeedback provides an external display of internal body reactions such as heart rate. This external mirroring increases the sensitivity for the interoceptive body sensations that provide subjective representations of body reactions. Importantly, too, it also enables subjects to gain a degree of voluntary control over their body reactions.

Parents' affect mirroring provides a kind of natural biofeedback training for infants. Initially, infants may experience emotional expressive and body reactions only as all-encompassing, unfocused proprioceptive and interoceptive sensations. Through affect mirroring, caregivers provide a contingent external representation of the infant's expressive reactions. Infants are preadapted to discover this contingency between their internal sensations and the external expression representation. Succinct affect mirroring of expression signs and the temporal contingency between the child's own subjective feelings and mirrored expression is repeated, day in, day out, for months. Progressively infants attribute the expression sign mirrored by the caregiver no longer to the caregiver but to the self and start to interpret it as an expression sign for their own current feeling (Gergely and Watson 1999). In this process, infants progressively develop a sensitivity for the proprioceptive and interoceptive sensations that correspond to their own emotion-specific expressive and body reactions. They learn to use expression signs to group those sensations in an emotion-specific manner and to relate them to their own corresponding causal attributions and coping actions. This process makes infants aware of their feelings. Unmediated, unfocused body sensations become conscious, cause-focused feelings.

In affect mirroring, both awareness of others' feelings and awareness of one's own feelings emerge. Once feelings have become conscious, they make it possible to adjust expressive and body reactions to fit one's motives, and they also take a step farther toward reflective emotion regulation. Accordingly, becoming aware of feelings not only emerges from the symbolic use of speech signs and during verbal symbol formation but is present much earlier in the symbolic use of expression signs (Malatesta and Izard 1984).

The symbolization of expression signs precedes and prepares the infant for language acquisition; the development and use of expression symbols prepare the infant for the acquisition and use of speech symbols. Genuine self-regulation of emotion becomes possible only through the acquisition of language. This is because genuine self-regulation requires the capacity to anticipate future emotion-related situations, plan to respond to them in certain ways, and resist alternative, possibly more appealing, action possibilities. This clearly requires symbolic thought.

Speech acquisition begins in infant protospeech (see chap. 5) but develops quickly up to the beginning of year three through two apparent stages; these are the stage of emotional and nonemotional expression independence and the stage of linguistic action appeal (Holodynski and Friedlmeier 2006, 116–117). In the first stage, regulation of emotional and nonemotional responses occurs independently. Infants use the first speech signs almost exclusively in emotion-neutral situations. As soon as the situation becomes emotionally charged, interaction returns to expression sign regulation (Bloom 1993). Infants in Bloom's study uttered their first word at 12.8 months, and their vocabulary spurt occurred at 19.2 months. Only at the time of the vocabulary spurt did toddlers begin to include speech in emotional expression. This indicates the integration of emotional and nonemotional action self-regulation.

As noted, first words are emotion-neutral and are used to show caregivers objects of interest. Infants show the object to the caregiver and utter a more or less accurate word for it. Caregivers repeat (mirror) the word, emphasizing correct pronunciation (exaggeratedly succinct) and demonstrating its relation to function, another object, and so on, by pretending to perform a typical action for it (e.g., digging with a spade) or uttering a typical sound for it (e.g., dogs go bow-wow). The almost exclusively declarative use of words in toddlers and the mirroring response of caregivers ensure that the two partners share a common attentional focus. This impulse to share is uniquely human; I return to discuss it in chapter 5.

The second developmental task that infants and toddlers need to master is the use of words as action appeals toward others. This is not to share an attentional focus but to use words to direct others instrumentally. Once the toddler starts to use words in this way, caregivers respond by doing the same. Toddlers' first use of words, however, is limited. They can only compel action. Toddlers from one to two years can comply with requests and instructions to perform simple acts, such as clapping hands, but cannot regulate or inhibit such acts. When asked to squeeze a ball when a light comes on, toddlers at this age start to squeeze immediately and repeat the squeezing. Even commands not to squeeze trigger the squeeze reaction (Luria 1961, 1980). Speech at this age possesses only an action-triggering function, irrespective of any specific meaning their instructions might convey. A "tiresome learning process" (Holodynski and Friedlmeier 2006, 118) is required before meaningful content of an utterance becomes effective in action and self-instruction can be used for one's own self-regulation.

In a further series of experiments, Luria found that three- to four-year-olds can reduce the number of errors when following simple self-instructions like "squeeze" and "don't squeeze," but found that, as with the two-year-olds, "don't squeeze" led to further squeezing. Children at this age therefore can initiate action by self-instruction but still not regulate or inhibit it. Not until five to six years does the meaningful content of an utterance become operational in action regulation.

4.2.1 The Emergence of Self-Evaluative Emotions (SEEs)

During the second year, children learn to consciously differentiate between themselves and others. They become aware of themselves as acting agents with their own intentions and expectations and are able to distinguish these from the expectations and intentions of others. This ability manifests in the use of the words *I*, *me*, and the child's own name and heralds what Stern (1985) terms "the birth of the verbal self" (see Holodynski and Friedlmeier 2006, 134).

To produce effects oneself becomes a strong motive, but the effects are always object directed rather than aimed at eliciting parental reaction.

Positive social evaluation is not an incentive for children at this stage, but it is used by parents to encourage its appreciation. Parents praise good behavior in exaggerated terms and communicate exaggerated signs of pride. They react not only with smiles and joyous vocalizations but also with exaggerated signs of admiration and pride: they puff out their chests and urge their children to do similarly, to stand up tall and feel grown-up. Their motor mimicry ability then allows children to attune to these expression signs through proprioceptive feedback and feel or "catch" the pride. The parental response thus stimulates a socially evaluative form of effect-related joy in the child, which is represented in expression signs of broad smiles, standing tall, and making celebratory gestures (K. Barrett 1995). The same sort of process occurs with the other self-evaluative emotions (SEEs). It is also the process through which children develop the ability to evaluate their own reactions in light of such norms, attempt to control norm-violating emotions, and divert reactions into the appropriate channels. This development is scaffolded (Scaffolding 1) by the increasing use of language in both caregivers and children in emotional regulation. Parents increasingly require children to request assistance verbally rather than just whine and hold out their hands, and children increasingly use verbal instructions and requests to others and to themselves. Children eventually learn to use SEEs to assess their behavior in relation to their culture's norms; they feel pride when their behavior accords with such norms, shame when it does not, and indignation when other people disregard them (K. Barrett 1995).

By four to five years, children are able to use symbolic strategies in emotion regulation. They reinterpret situations, use imagery, and organize their goals according to priority. What all of this means is that intrapersonal regulation emerges through children doing for themselves what previously their caregivers did for them. This process is supported by the parents' increasing refusal to do things for their children; they tell them to do it for themselves but often include instructions on what to do: "Do it yourself, like this …" This shows another aspect of sensitive parenting; it shows that during this early period of increasing intrapersonal regulation, children still need the guidance and support of a competent other.

From about four to six years, socially focused reactions to emotional episodes (e.g., seeking mother) decline, to be replaced increasingly by problem-focused (e.g., working harder, waiting for assistance) and emotion-focused (e.g., making fun of the situation, controlling expression, and shifting attention) reactions. (However, it is the case that extremely intense emotional episodes require social support throughout the life span.)

Preschoolers initially show pride or shame only in social interaction with adults; they show only effect-oriented joy or frustration when alone (Holodynski and Oerter 2002). Only as they progress through elementary school do they start to react with pride and shame just as adults do. Young children, accordingly, still require social interactions in which an adult represents norms through her behavior if children are to experience pride after a success rather than simply effect-related joy. The same is true for shame after a failure or a moral transgression, rather than just effect-related disappointment or frustration.

What this indicates is that the self-evaluative emotions also emerge through the interplay of caregiver affect mirroring and children's motor mimicry. In coregulation with their child, caregivers establish emotion-specific contingencies between cause, social evaluation, expression reactions, and actions. Caregivers mirror child expressive reactions in their own expressions and present expression signs in their own behavior as models. This affect mirroring coincides with children's sensitivity for contingencies and their ability to "pick up" the feelings of their partners through motor mimicry.

4.2.2 Cultural Shaping of Emotions

> It is a species-specific characteristic of human beings to use socially created artefacts (in the form of tools and symbols) to reshape their naturally given environment within a process of social coordination, and to pass on this culture to the following generations. These, in turn, adopt the artefacts and apply them to regulate their activity and to maintain social and cultural life.
>
> —Holodynski and Friedlmeier (2006, 182)

Each culture is characterized by ethnotheories that are often implicit but are "shared set(s) of beliefs, attitudes, norms, values and behaviour organized around a central theme and found among speakers of one language, in one time period and in one geographic region" (Triandis 1995, 443). Clearly, one such central theme in every human culture is child rearing, and ethnotheories clearly do inform parental child-rearing practices from the beginning (Trommsdorff and Kornadt 2003); the interpretation of the still-unfocused precursor emotions always occurs in light of their emotion-related ethnotheories. Two types of ethnotheory interact to inform child-rearing practices; these are ethnotheories that encapsulate the general norms and values for the appropriate conduct of human social life, which have normative consequences for the emotions; and ethnotheories that relate specifically to emotions. These are transmitted intergenerationally through parental (and school-related) socialization practices. Both types of ethnotheory inform the different intuitive parenting behaviors of caregivers in terms of coregulation, affect mirroring, and responding to infant feedback (see the previous section).

In terms of general norms and values, in a series of cross-cultural surveys, Hofstede (2000) found five dimensions of person-world relations that distinguish between cultures: (i) individualism–collectivism, (ii) power-distance, (iii) masculinity, (iv) uncertainty avoidance, and (v) long-term versus short-term orientation. The most important dimension in emotional ontogenesis is individualism–collectivism. The dimension of individualism or collectivism refers to the extent to which a particular culture favors individual goals and values compared with those of the larger reference group (Triandis 1995). Individualistic cultures emphasize the personal significance of feeling in emotion; they view emotions as legitimate and important and view emotional expression as both healthy and positive. In contrast, collectivist cultures view emotions as socially situated, that is, as a relation between a person and her environment; consistent with this view, the social situation in which emotion occurs is seen as important, as is the control of emotion, but personal feeling is not. Western cultures emphasize individual independence, encourage emotional expression, and expect self-confident and competitive behavior and are disappointed with socially inhibited behavior (Holodynski and Friedlmeier 2006, 194), and the opposite is true of collectivist cultures (Trommsdorff and Friedlmeier 2004). In China, for instance, even four-year-olds are expected to control their emotional reactions (Ho 1986). Culture-specific ethnotheories, in particular, inform affect-mirroring behaviors and coregulation strategies.

Affect mirroring focuses on single emotion types, and the less affect mirroring of a particular emotion type, the less likely it is that children will construct a categorically discernible feeling state from their emotional reaction. This occurs when caregivers either ignore or counter with distraction strategies emotional reactions construed as negative, such as anger. In a study of twenty-four Japanese, Chinese, and American children between eleven months and two years, Camras et al. (1998) found that American children were more emotionally expressive in general and expressed anger freely, whereas Japanese children reacted to an anger-eliciting situation with sadness rather than with anger. In addition, the researchers found that both Japanese and Chinese mothers mirror affect expressions less than American mothers and prefer never to show anger. They even try to prevent their infants from screaming and crying. In a study of Japanese and American mothers, Rothbaum and Morelli (2005) found that Japanese mothers intervene proactively to prevent negative emotion, whereas American mothers tend to intervene reactively to deal with an actual emotion episode. Japanese mothers limit infants' exposure to environmental stimuli, limit their own introduction of novel stimuli in their interactions with the infant, and also limit their attempts to manipulate infants' attention. Interestingly, both sets of mothers agreed that proactive intervention would promote accommodative behaviors and the opposite would be true of reactive intervention. Of course, proactive intervention limits opportunities for emotional exploration and the development of independence and intrapersonal emotional regulation. This results in the later attainment of Japanese children's intrapersonal emotional regulation (Holodynski and Friedlmeier 2006). Differences in the expression of shame and anger in collectivist cultures of, for example, many Asian cultures compared with those of typically individualistic Western cultures illustrates this influence (Mascolo and Griffin 1998).

Anger is a reaction to the hindrance or prevention of one's goal attainment and leads to readiness to act to overcome these obstructions. By contrast, shame reflects one's own violation of social norms and readiness to subordinate oneself to these norms and leads to readiness to isolate oneself and comply with such norms in future. As would be expected, anger is less desirable in collectivist cultures because it is socially disruptive, and shame is viewed positively because it promotes social harmony; the opposite is the case in individualistic cultures. In individualistic cultures, anger is viewed positively, provided that it is expressed in socially acceptable ways, since it is seen to promote independence and self-expression; shame is viewed negatively because it undermines personal self-esteem, but again, the opposite is true of collectivist cultures. In Asian cultures, a person is expected to adapt to the environmental situation by controlling the psychological effects of the emotion on herself and using calming, distracting, and reappraisal to control her reactions. In contrast, in Western cultures, persons are expected to use problem-focused strategies to alter the social or physical environment to achieve their goals.

In a series of studies using children from the United States and Germany and China and Japan, Holodynski and Friedlmeier found clear evidence that both at home and later at school children learn to express themselves according to the individualism or collectivism of their host culture. In Germany and the United States, children are encouraged to express themselves emotionally as a reflection of an independent self-concept, whereas in China and Japan, children are expected to control their emotions and conform to group norms as a reflection of an interdependent self-concept (Holodynski and Friedlmeier 2006, 209–210). This is at least partly the result of deliberately limiting exposure to potentially emotional situations and limiting parental affect mirroring of emotions; Asian children typically display less emotion than Western children.

There are four important points to note at this juncture; I begin with the most obvious. The first is that affect expressions are the basic elements of human emotionality from which BEs and HCEs subsequently develop. They are clearly biologically and psychologically primitive (Ortony and Turner 1990) and can be at least roughly identified with neuroanatomical brain regions (Panksepp 2000). In addition, they are clearly homeostatic property cluster kinds (see chap. 2). Homeostatic mechanisms, such as very loud noises, bring about the co-occurrence of their properties, for example, facial expression and startle. Given this, it appears that emotions are a natural kind.

The second point is that the ability to emote properly at all, as a human being, is a function of linguistically mediated social relationship. When everything is functioning normally, that is, ideally as it evolved to (but seldom does) in terms of neonate and caregiver input and output, the relationship provides the necessary and sufficient conditions to initiate the development of full human emotionality from a set of culturally universal and monomorphic emotional precursors. This emotionality includes the identification of culturally relevant emotions, their culturally relevant elicitors, what they are called or labeled in the emoter's particular culture, and what the culturally appropriate response is. What this clearly implies is that (i) culture and language make a serious constitutive contribution to the development of human emotions, including basic emotions; and (ii) the achievement of intrapersonal emotional regulation is the result of the cultural and generational inheritance of the relevant ethnotheories.

The third important point to note is the synchronized operations of infant and caregiver throughout emotional ontogenesis. The affect expressions of early neonatal life are producers of unfocused, undifferentiated signs or expressions of threshold deficits with a combination of both indicative ("something's wrong") and imperative ("do something!") functions. These trigger the coarse-grained interpreter devices with which caregivers are equipped to discover what is causing their infant to (precursor) emote and what to do about it. The repeated exercise of such search processes, complemented by relevant "baby talk" (Holodynski and Friedlmeier 2006), fine-tunes these interpreter devices to the particular idiosyncrasies of their infants' needs, expressions, and satisfaction criteria ("Oh, Amy sad," accompanied by exaggerated frown and downturned mouth, "Amy hungry?" "Oh, Amy got windy-pops?" "Oh, Amy wet diaper," "Mummy make Amy comfy and dry," and so on). My point is that the underlying neural circuits and the differential feelings, causes, and satisfaction criteria associated with basic emotions mature or develop from affect expressions through this process. This must be the case. First, absence or interruption of close social relationships disrupts the development of normal mental capacities, including emotional capacities in all primates (see sec. 4.2). Second, and if only because of the Muggle constraint (Wheeler 2005, 4–5), that is, that all magical explanations are precluded, such capacities are necessarily metaphysically realized in realizer systems that include neural substrata.

This is entirely consistent with what developmental biologists have repeatedly demonstrated, that is, that the development of any trait or characteristic is a consequence of coactions between complex molecular interactions within and across cells and the nature of the physical, biological, and social environments through which individuals pass during development (Lickliter 2008). Coaction refers to the results of the relationship of interacting components in any developmental system, not the additive effects of the components themselves. Importantly, of course, such coactions inevitably involve genes.

Biological scientists now recognize that one of the clearest links between gene activation in individual neurons and experience is that of immediate early genes (IEGs) and sensory stimulation (Johnston 2008; Johnston and

Edwards 2002). Auditory, visual, tactile, and olfactory stimulation almost always results in the activation, within minutes, of one or more IEGs. These genes produce transcription factors to control the production of proteins by other genes, and these proteins construct and reshape neural circuits. The genetic activity is not restricted to early development; it is, rather, ever present during the life of human and nonhuman animals. This explains why animals, including human animals, who are raised in total or relative social isolation fail to develop normal cognitive and emotional capacities; without the requisite auditory, visual, tactile, and olfactory stimulation, there will be no activation of IEGs and no construction and reshaping of neural circuits. This clearly supports the claim that realizer systems are metaphysically context sensitive all the way down. It also indicates that emotionality (and cognition more generally) is social in a much deeper sense than formerly construed. The neural circuits that underpin assistance-soliciting and assistance-producing devices of BEs and HCEs are developed and fine-tuned in the human neonate and child in response to caregiver responses to emotion precursors and differentiated (or differentiating) emotions. In addition, the reshaping of caregiver circuits that underpin the transition from coarse-grained to fine-grained assistance-providing devices results from the feedback they receive from their own search-identify-correct responses to the affect expressions of their infants and differentiated (or differentiating) emotions of their children.

Of course, one could object that human neonates do not require a close, linguistically mediated social relationship in which to develop their emotions because some children, such as the wolf boy of Aveyron, seem to have developed basic emotions in the absence of such a relationship. We should note a distinction, however, between human emotionality and emotionality in other species. This is not to deny that wolves emote in ways relevant to wolf lifeways; one would expect them, as social mammals, to emote in ways appropriate to the regulation of pack life. Human infants reared in close contact with wolves, therefore, would learn the emotions appropriate to pack life, and these emotions do seem to have analogues at least to the emotion precursors we pick out with the English words *fright, distress, interest,* and so on. This seems unremarkable; basic emotions and their precursors are underpinned by neural circuits in primitive areas of the brain that human beings share with other nonhuman species, and this could account for the similar "emotionality" apparent in wolves (and other nonhuman species). Human infants, just like all other nonhuman animals, can learn to respond emotionally to certain stimuli simply by association. In my account, however, they will not emote fully as human beings until they can semantically conceptualize the emotions they experience, their causes, and their appropriate responses.

This is an important point; as children develop and grow, they learn that an expanding range of stimuli can elicit basic emotional responses, and these typically do require semantic conceptualization. This is not to deny that subdoxastic concreta (e.g., slowly moving "speaking" heads, snake-shaped objects), music, and changes in facial expression and bodily orientation can elicit basic emotions, for clearly they can and do. It is, rather, merely to claim that semantic conceptualization is also an important elicitor. In cases such as blue exam books (Prinz 2004), for instance, and summonses from senior colleagues, the emoter-subject necessarily requires some grasp of the meaning, in terms of potential threat, of blue exam books and summonses from senior colleagues to experience a fear response. Nor is it to deny that BE responses can be automatic, for clearly here, too, they can be. It is, rather, merely to claim that some degree of top-down elicitation and regulation is possible in some situations.

Emotions are stimulus dependent, which means that they are constrained by their inputs. This is consistent, however, with the possibility that top-down processes can significantly affect the operation of perceptual systems. Bottom-up inputs trump top-down inputs when the two conflict (Prinz 2004). This might be because they activate phylogenetically ancient sensory thalamo-amygdala circuits in the midbrain much more speedily than the more phylogenetically recent sensory thalamo-cortico-amygdala circuits. When there is no bottom-up/top-down conflict, top-down inputs operate efficiently.

The fourth important point is that the entire human emotional ontogenetic process involves the progressive fine-tuning of the relevant developmental mechanisms, and this occurs through progressively fine-grained mutual modulation of neonates' assistance-soliciting and caregivers' assistance-providing interpreter mechanisms. It is through these mutual modulation processes that neonate and caregiver become intimately acquainted and attached, that is, develop the special social relationship on which human emotional ontogenesis partly but crucially depends. In addition, such mutual modulation simultaneously results in the progressive development, maturation, and connection of the relevant neural substrata on which, again, human emotionality partly but crucially depends. Both developing social relationship and neural substrata enable further, progressively synchronized mutual modulation of neonate and caregiver mechanisms. In short, the special social relationship and neural substrata are both causes and effects of increasingly synchronized mutual modulation. The degree of complexity, complementarity, and integration is such that without the coaction of all neural and extraneural components involved, human emotionality would fail to develop or to develop normally.

In sum, emotional ontogenesis provides a compelling example of deep functional integration and CT. Emotional ontogenesis, as noted previously, is crucially dependent on the close, linguistically mediated relationship of neonate and caregiver. The preadaptations with which both partners are genetically endowed, that is, neonates' emotion precursors (affect expressions), precursor emotion regulation strategies, and sensorimotor competences, and caregivers' intuitive parenting skills, coevolved precisely to promote such deep functional integration. The frequency and intensity of the visual, auditory, and tactile stimulation in the affect mirroring and baby talk commentary typical of intuitive parenting skills engage precisely with the precursor strategies and sensorimotor competences, especially motor mimicry of neonates. It is through the coactions of affect mirroring, baby talk commentary (caregiver), and motor mimicry and rewarding feedback, that is, cooing, smiling, visually fixating, and snuggling up (neonate), that the coarse-grained search-identify-correct interpreter devices of caregivers are progressively fine-tuned and the coarse-grained affect expressions in the neonate are progressively fine-tuned into the producer and interpreter devices of basic emotions. It is through this interdependent (deeply functionally integrated) process that contingencies between emotion expression, feeling, and responses are synchronically constructed in both partners because the requisite neural substrata are being developed, matured, and reshaped. In addition, subsequently, precisely the same interdependent (deeply functionally integrated) process of affect mirroring and verbal commentary, including instructions (caregivers), and motor mimicry and verbal commentary, including instructions (children), promotes the development of the higher sentiments (Holodynski and Friedlmeier 2006). The level of deep functional integration of both neonatal and caregiver neural and corporeal resources with external sociocultural resources in emotional ontogenesis is such that they constitute a genuinely extended cognitive system by affecting the system as a whole, producing effects that result only from the system functioning as a whole. What this clearly implies is that component structures and processes in human emotional ontogenesis operating across two socially embedded brains are deeply metaphysically interdependent. In light of this, it is a mistake to construe causality and constitution as mutually exclusive relations; causal relations can be, and in this case very clearly are, constituents of larger, more complex structures and processes. In addition, the level of DFI in emotional ontogenesis is such that it is a mistake to maintain that cognition always extends from brain to world. The constitution of human emotional ontogenesis in neonates and children and the fine-tuning of caregiver intuitive parenting skills are a result of mutual, synchronized modulation of causal processes in and across (at least) two socially integrated craniums. In this account, human emotional ontogenesis extends from the brains of people in the external environment and the sociocultural products into which they are deeply integrated into other people's developing brains. What this clearly implies is that emotional ontogenesis is a world-to-brain transcranial achievement.

In the externalist account for which I argue, both wide and radically wide realizations are involved in human emotionality. I have argued that the intrinsic physical states of the individual bearer of emotional states (and affect expressive states) are necessary but not sufficient for the metaphysical realization of human emotions. The total realizer systems, in terms of both emotional ontogenesis and adult emotionality, extend beyond the skin and skull of the emoter-subject into the sociocultural environment into which she is deeply embedded. The metaphysical realization of human emotionality, therefore, is never merely entity bound or individualistic.

Human emotional ontogenesis is radically widely realized; its core realizer is the close, complexly interdependent, linguistically mediated social relationship of neonate or infant and caregiver, which clearly extends beyond the physical boundary of the (precursor) emoting infant. This relationship provides, inter alia, the auditory, visual, tactile, and olfactory stimulation required to activate the immediate early genes (IEGs), which progressively construct the neural circuits that eventually underpin adult emotionality. These circuits account for both the automatic processing of certain emotion-related sensory-perceptual stimuli and semantic and conceptual processing of sensory-perceptual and linguistic and symbolic stimuli. In short, this relationship is critical to the gradual development, maturation, and connection of emotion-related neural circuits, including the amygdala, which underpin the progressive shift from interpersonally regulated juvenile emotionality to intrapersonally regulated adult emotionality. In the absence or disturbance of this relationship, human infants will fail to develop into full-blown human emoters, as the empirical evidence bears witness. (I deal with the changing locus of control in developmental interaction sequences in chap. 5.)

The total realizer system of human emotional ontogenesis, therefore, which includes core and noncore partial realizers, is metaphysically realized in a range of heterogeneous but complementary internal physical resources (e.g., neural substrata, IEGs, optic nerve, facial musculature) and external sociocultural resources (e.g., language, emotional ethnotheories).

In full adult emotionality, the core realizer is the amygdala; it is, as indicated previously, the part of the total realizer system most readily identifiable as producing and sustaining full-blown adult emotionality. In this case, the core realizer does not extend beyond the physical body of the emoter-subject, but the total realizer system does. Adult emoters understand the meaning of the various emotion elicitors to which they are subject, the subjective feelings that they engender, and the culturally appropriate responses to them, and these depend on the sociocultural and linguistic traditions into which they are also deeply embedded. Thus adult emotionality is also realized in a heterogeneous but complementary mix of internal physical resources (e.g., amygdala, cortex, facial musculature) and external sociocultural resources (e.g., language, cultural ethnotheories).

4.3 Conclusion

I noted in the introduction that intracranialists claim that extraorganismal resources can be deeply functionally integrated with intraorganismal resources, yet deny that DFI can provide evidence of CT. I also noted that transcranialists claim that DFI hinges on the extent and complexity of functional complementarity and integration; thus DFI is a matter of degree. Deep functional integration with CT, I have claimed, is achieved when the internal and external components in a cognitive system function synchronously as a whole, producing results only by functioning synchronously as a whole. However, since DFI is a matter of degree, the deeper the functional integration of components, the more the system will function synchronously as a whole, and the more it will produce the effects that result only from the system functioning synchronously as a whole. Inevitably, therefore, there will be cases of functional integration that are relatively weaker and relatively stronger in terms of complementarity, complexity, and integration, and claims at the weaker ends of these dimensions may well appear strained. Claims to CT at the stronger ends, however, will appear natural and justified. This will be particularly true when such cases are described in the kind of full and rich detail (Sutton et al. 2010) as the description of human emotional ontogenesis given in this chapter. What this implies is that intracranialists' putative explicit acceptance of DFI while denying CT can only pertain to cases that are relatively weaker in terms of complementarity, complexity, and integration.[3]

5 Out of the Mouths of Babes and Sucklings

5.1 Introduction

In chapter 1, I identified two forms that biological constraints can take. These were sensorimotor competencies and species-typical activity patterns, and in chapter 4, I identified certain sensorimotor competencies with which human neonates are preadapted and applied these particularly to the learning of human emotionality. In this chapter, I identify early species-typical activity patterns that are present at birth or develop in early life, and show how they provide the basis for the acquisition of language. I argue that the linguistic environment into which human neonates are born and the complementary repertoires of constraints (species-typical behavior patterns) with which neonates and their primary caregivers are biologically equipped provide the necessary and sufficient conditions for the construction of what appears to be a two-stage process of language acquisition, incorporating first a pragmatic foundation and subsequently a lexicon. These constraint repertoires are as deeply functionally integrated as those that underpin emotional ontogenesis and serve, again, to help construct the close, linguistically mediated social relationship within which both emotions and language develop. What this implies is that the core realizer of language acquisition is the same close, linguistically mediated social relationship as that of emotional ontogenesis and that language acquisition is also radically widely realized.

I argue that the developmental mechanism is, again, (i) initially, automatic, unthinking, exaggeratedly succinct parental mirroring of infants' protowords, pitched to match their perceptual capacities and structured to establish the prelexical pragmatic context; and when the child signals her readiness or interest, (ii) a deliberately pedagogic approach to language learning. In light of these analyses, I also argue that a rich, innate language acquisition device is probably superfluous; very limited innate repertoires of constraints and the social environment probably provide all that is required for language development in normal children. I argue, further, that the learning of both language and emotion begins in turn taking, which is established through suckling, either breast or bottle, and that the constant daily repetition of simple vocalizations and gestures is fueled by the same endogenous opioids and prosocial neuropeptides as emotional ontogenesis.

In addition, in this chapter, I include a similar level of detail as that in chapter 4 to (i) preclude any intracranialist denial of genuinely transcranial processes operating in language development, and (ii) replay my rebuttal, albeit with details from language development, of intracranialists' claim that transcranialists commit a coupling-constitution fallacy. I conclude the chapter by describing the opioids and prosocial neuropeptides that fuel neonate-caregiver interaction.

5.2 Species-Typical Activity Patterns

Behaviors that are present at birth or emerge in the early months include reach-and-grasp movements (Fischer and Bidell 1991); early lip and tongue movements, or "prespeech" (Trevarthen 1979); rhythmic supine kicking and hand waving (Thelen 1981); and, most important, burst-pause-burst suckling (Kaye 1982). These, I suggest, are the nonemotional or emotion-neutral analogues of the unfocused crying and motor unrest and cooing, smiling, and snuggling up of emotional expression (see chap. 4). Primary caregivers are aware that such unfocused activities are not emotional reactions simply because their infants are manifestly not distressed or very happy, but caregivers correctly assume that these activities are minimally intentional insofar as they are meant to attract parental attention. In the same way that mother is preadapted to search-identify-correct (distress) and prolong and intensify (happiness) the cause of her infant's emotional reaction, she is preadapted to search-identify-satisfy her infant's nonemotional wishes and demands. In exactly the same way that she explores the possible causes of emotional distress or pleasure (e.g., by checking to see if the baby's diaper is wet, watching to see if she sucks her fingers, shivers, etc.), the mother explores her infant's possible wishes (e.g., by talking, picking up, playing peekaboo). What this implies is that the search-identify-correct or prolong-and-intensify responses to emotional expression and the search-identify-satisfy responses to nonemotional species-typical behavior patterns are both forms that intuitive parenting can take. Further, because both correcting or prolonging and satisfying wishes and demands are both means of satisfying infant need, intuitive parenting more generally involves search-identify-satisfy responses. In addition, again, similar to the role of affect expressions in emotional development, if hand waving results in being picked up and cuddled, and the infant responds with appropriately positive feedback, such as smiling, cooing, and cuddling up, the mother will repeat her lifting behavior the next time the infant waves her hand so as to elicit the same response. The same holds true for reach and grasp and supine kicking and prespeech. These are also interpreted as nonemotional assistance-soliciting signals that elicit in caregivers complementary, assistance-producing behaviors. I noted in chapter 4 that caregivers actively seek opportunities to interact with their babies, and the assistance-soliciting species-typical behavior patterns provide some of these.

There should be one important difference, however, in the physiological underpinnings of assistance-providing motivation in nonemotional assistance-soliciting situations. There is no need for a physiological stress reaction; the infant is manifestly not distressed. The assistance-providing motivation, therefore, is presumably a result of prosocial neuropeptide flooding at birth and during suckling. These prosocial neuropeptides excite the intraneural genetic products into neurogenesis and neural connection in neonate and caregiver. These new or more elaborate neural circuits underpin progressively fine-tuned assistance-soliciting and assistance-providing devices in neonates and caregivers, which, in turn, underpin more fine-tuned and elaborate forms of mother-infant interaction. What this implies is that the species-typical behavior patterns and the responses they elicit, or the deep functional integration of complementary repertoires of neonatal species-typical behavior patterns and parental responses, are as instrumental in the development of the very close, linguistically mediated social relationship of neonate and caregiver as the emotionally focused complementary repertoires of neonate and caregiver. This relationship therefore provides the context for the continuing development of more complex forms of emotional and nonemotional behavior.

5.3 Turn Taking in Human Development

These species-typical behavior patterns are important, for they are all instrumental in the development of turn taking with the mother. Her responses, qua responses, follow infant activity and result in interactive sequences of turn taking (Bloom 1991; Bloom and Tinker 2001). A mother's response to a pause in suckling (breast or bottle) is probably the most important in this respect; according to Hendriks-Jansen (1996, 253), it is the foundation for the "conversational dialogue" that develops between mother and infant. I agree with Hendriks-Jansen that this is one of the functions of turn taking, but not its main function; see below.

Human suckling is unique in that it is punctuated naturally by random pauses that may be due to the anatomical changes in the upper respiratory and digestive tracts that enable speech. Mothers notice these pauses because suckling is a distinctly rhythmic activity, and because such pauses could lead to malnourishment if allowed to persist, they are interpreted as signs of laziness or disinterest ("Something's wrong! Do something!") or even of a desire to play rather than suckle ("Something's right! Do something!"). Mothers are preadapted to respond by jiggling (M. Bateson 1979), which is meant to encourage the infant to return to suckling.

Interestingly, it is not the jiggling but its cessation that results in the resumption of suckling; mothers, however, are not aware of this. What this means is that jiggling per se has no nutritional value; this further implies that it was selected for another function. Pause-jiggle sequences are clearly minimally intentional (they are produced for consumption or use by another device) and thus coevolved to establish turn taking. Feeding thus becomes a process of turn taking: suckle-pause-jiggle (pause)-suckle-pause-jiggle (pause)-suckle, and so on. It was Kaye (1982, 188) who suggested that this turn taking resembles ordinary human conversation, where one speaker speaks and then pauses, and during or as a result of the pause, her interlocutor then speaks.

My view is that both Kaye and Hendriks-Jansen underestimate the proper function of turn taking, which, they suggest, provides the platform for language acquisition. Suckling is probably the very first activity engaged in by human infants, and if suckle-pause-jiggle coevolved to enable turn taking to develop, then turn taking is one of the first activities to develop in every neonatal life. This implies that it is just as important in emotional ontogenesis as in language acquisition, and for the same reason. It provides essential opportunities for infants to mimic and practice behavior and for caregivers to provide feedback by means of, for instance, exaggerated affect mirroring and baby talk commentary and, as the child grows, to correct pronunciation and give instructions and advice. Turn taking, in this account, is not primarily to enable conversational dialogue per se: it is, rather, to enable learning of the essential skills of emotional control, language, and symbolic thinking. A crucial implication of this is that human children learn to speak through the same process through which they learn to emote, that is, through motor mimicry and tediously repetitive practice, day in, day out, with interactive caregivers.

The other patterns of activity listed earlier are apparently initially fired off at random; the infant coos, kicks, and waves her arms, and the mother correctly interprets these as primitive intentional acts to which she needs to learn to respond appropriately. Each activity evolved initially to exercise and develop the neuromuscular apparatus that will underpin more complex motor and speech acts, but was subsequently selected (together with maternal responses) to provide a turn in a "conversation" where the infant moves, then pauses, then the caregiver fills in the gaps, then pauses to allow her infant to respond, and so on. That this is the case is evidenced by the irrelevance of some of these activities (e.g., suckling pauses) to physiological well-being, and the complete disappearance of others in the early months of life.

The caregiver thus allows herself to be paced by the infant, and all of this reinforces the interactive turn taking already established through feeding. The important point is that the mother responds to these expressions as minimally intentional, but she is the only one of the partners who can attribute meaning. There is no substantive content to maternal-infant "dialogue" in the early months, but it is foundational in establishing the pragmatics of conversation that will enable the emergence of meaning for the infant in due course. In addition, the supine kicking, hand waving, and reach-and-grasp movements are developed and refined through highly repetitive motor sequences often scaffolded (Scaffolding 1) by the provision of bright, interesting toys (see sec. 5.6).

Turn taking provides the basis for a number of important developments over the next few months:

- The infant discovers which activities elicit which responses from caregivers.
- She develops routine sequences of interaction that are triggered predictably.
- These provide a context for the development of mutual expectations.

5.4 The Emergence of Joint Attention

I noted earlier that by the age of six weeks, mother and infant are communicating intensively face-to-face and engaging in mutual gazing (Chappell and Sander 1979; von Hofsten and Rosander 2013). This is enabled by the increasing opportunities for interaction afforded by increasing neonatal alertness during mundane nurturing and cleansing activities in the early days and weeks. In addition, early face perception is enabled because it matches the operation of human sensory, perceptual, and cognitive mechanisms (Bushnell 2013). Mother places herself directly in line with the infant's eyes and at a distance of 20 to 24 centimeters, gently moving her face back and forward and speaking to the infant; the infant is preferentially disposed to follow speaking, slowly moving heads (Simion, Regolin, and Bulf 2008; Troje 2013), and the infant's eyes often fixate on the maternal face. At this age, the infant's visual system, although poorly developed, is sufficiently developed to allow the infant to follow movement, and the mother's actions ensure that what she sees is her face. This gives the mother the impression that her infant is showing interest in her, and thus she interprets the gaze as intentional.

This interpretation of intentionality is reinforced from about one month, when the infant begins to fixate on the mother's face for extended periods; again, she interprets the gaze as intentional, which she finds rewarding. In turn, this enables the mutual attention and shared emotion of affect attunement that characterizes Trevarthen's (1979) primary intersubjectivity. At about six to eight weeks, the infant can track more fine-grained features in her visual field. This enables her to attend to the mother's features and changing facial expressions, again increasing the impression of intentional response.

At four months, the infant's visual behavior changes noticeably; having established intense face-to-face contact with mother, she begins to look away. Although no evidence at this stage suggests that the infant is foveating on particular objects, mother assumes that this is the case. The infant, however, may perform a random reach-and-grasp movement, possibly coincidentally with the direction of her gaze, and if an object is present (and the mother scaffolds the environment to ensure that there is), the mother interprets the movement as intentional. She assumes that the infant is expressing a desire for the object, which the mother then hands to the infant. By so doing, the mother incorporates a species-typical activity pattern into an interactional sequence that is about something. "With the aid of the dynamic scaffolding supplied by the mother, the infant thus performs intentional acts (acts directed towards objects and about objects) long before he is capable of intentional thought" (Hendriks-Jansen 1996, 273). This is because the mother is preadapted to behave with regularity and thus structure the interaction in ways that become familiar to the infant. After several responses of handing over the object, whether or not the infant actually signaled that she wanted it, the infant learns how to behave so as to have objects passed to her. This kind of interaction results, at about six months, in pointing emerging naturally as a communication gesture in infant-mother dialogue.

The infant points to an object in the environment, and the mother interprets the gesture in folk psychological terms, that is, the infant desires the object. This is imperative pointing and results in the child being given what she wants. However, two other forms of pointing emerge after this; these are informative and declarative-indicative pointing, and they have different purposes. Imperative pointing (rather obviously) indicates what the child wants; informative pointing assists others by providing information that they might need (Aktar and Tomasello 1998); and declarative pointing is undertaken simply to share interest with another person (Liskowski et al. 2006; Liskowski et al. 2004). Declarative pointing is undertaken simply because it is rewarding (Wyman and Tomasello 2007, 233), and it appears at around one year.

It is around this time that toddlers hold out objects to caregivers simply to show them, thus bringing the object into common view. They also begin to direct the attention of others through indicative gestures, in particular by pointing to objects and events in the environment. This indicates a transition to secondary intersubjectivity (Trevarthen and Hubley 1978), where infant and mother can focus on an external object rather than just on each other. Two other behaviors emerge around this time. The first is that children show concern for others in distress and occasionally try to comfort them. Presumably this is learned through imitation of caregivers; see chapter 4. The second is the coordinated activity of toddler and adult in highly ritualized games, such as peekaboo or rolling a ball back and forth. These ritualized games are scaffolded (Scaffolding 1) by interacting adults. Even at this age, toddlers understand that the roles are interdependent; children prompt adults to resume playing if they stop. By 15 to 18 months, noticeably more frequent joint interaction emerges in free play situations, although still ritualized to some extent, which involve child, adult, and object or event. The child can both follow the adult's lead and actively direct her attention (Wyman and Tomasello 2007, 470). By two years, however, interaction is occurring where child and adult cooperate much more on spontaneous rather than ritualized projects or problem-solving tasks. Should the adult cease interacting, again the child invites resumption. Thus, by two years, the child understands her role and her partner's role as interconnected in joint activity (471). She also understands that she herself and those with whom she interacts can attend to the same object and, in addition, can monitor each other's attention to that object.

These pointing and showing behaviors provide evidence of an early capacity to construct and negotiate interactional sequences ("dialogues") with others, based on mutual attention. This enables children to focus jointly with others on objects and events that can become the object of a shared goal, such as building a tower with building blocks. And, of course, as toddlers learn their native language, they learn that words can be used in place of gestures as attention-directing devices (Tomasello 1999), such as "Dada!" By sixteen months, pointing is consistently accompanied by one word, often monosyllabic or reduplicated (e.g., *Mama, Dada, Nana*) and related or relevant to the primary interests of the child (M. M. Lewis 2013; Bloom and Tinker 2001).

A strong relationship exists between pointing and language development (Rowlands 2014). In a meta-analysis of language development from 25 studies involving 734 children between 1978 and 2009, Colonnesi et al. (2010) showed a predictive relationship between the two. They conclude that "the pointing gesture is a key joint attention behaviour involved in the acquisition of language" (352).

It is worth emphasizing that the key to shared collaborative activity—and, crucially, this includes communicative activity—is joint attention. It is unique to the human species (Butterworth 2013). A series of comparative studies of human toddlers and chimps of one to five years (Tomasello and Carpenter 2005; Warneken, Chen, and Tomasello 2006) discovered that chimps engage in no joint attentional behaviors at all. In addition, Tomasello and Carpenter (2005) found that chimps fail to develop any understanding of role and role reversal. Both sets of subjects were involved in a game requiring (minimal) cooperative interaction with a human experimenter, and although the chimps managed to imitate some of the experimenter's actions, there was no evidence that these actions were conceived of as roles in the sense that they were intended for the partner. Of particular interest was the fact that the chimps failed altogether to look at the experimenter partner. Wyman and Tomasello (2007) suggest that chimps can coordinate their actions with others but do not conceive of the interaction as composed of designated roles that are interdependent and therefore reversible. It is the unique ability of the human species to do this, and it is this on which cooperative activity depends.

According to Butterworth (2008, 2013), joint visual attention (JVA) emerges when both participants engage focally with an object, and it commences in primary intersubjectivity (see earlier). From about one month, the infant and mother attend focally to each other's gaze, and from about four months, there is some evidence of joint attention to objects, albeit of a fleeting and infrequent nature. Butterworth (2008, 214) terms this "fragile triadic JVA." This is followed, at about nine months, with "robust triadic JVA," which characterizes secondary intersubjectivity. This change is accompanied, from about eleven months, with canonical pointing or pointing proper.

In addition, comprehension of pointing precedes production by some two months.[1] The communicative intent of pointing is fully understood by twelve months, and the production of pointing begins around fourteen months. In one of their studies, Butterworth and Grover (1989) found that infants at six to nine months fixated on a pointing hand rather than the target object, and if they did fixate on the target, they did so only after initially fixating on the hand. By twelve months, however, they fixated almost immediately on the target object.

In terms of production of pointing, infants point only in the presence of a social partner, never when they are alone. In terms of both declarative and imperative pointing, moreover, infants check, sometimes repeatedly, to see if their partner is attending to the target (Grover 1988). In addition, at around 12 to 14 months, another developmental transition occurs. Pointing becomes dissociated from the object pointed at; when the mother tries to hand the object to the child, she refuses to take it. Pointing is now used extensively to draw attention to unfamiliar objects, objects in pictures, and even objects not in view. The child looks at mother while pointing and perhaps utters a word (although not recognizable at this stage). Pointing has become a general-purpose means of expression.

When young children learn to communicate gesturally, they are learning that other people intend to direct their attention within a joint interactional sequence and that they can direct the attention of others in the same way. By fourteen months, toddlers attend to an adult's point as a communicative cue, can identify the referent of the point, and can infer the relevance of the point to their goal, say, finding a hidden toy (Behne, Carpenter, and Tomasello 2005). "They understand that the adult's motive is to help them and that she intends for them to attend to something. That is, meaning that they understand 'communicative intentions' (Grice 1957)" (Wyman and Tomasello 2007, 232). By fourteen months, therefore, human toddlers understand what ostensive communication is and its relevance to their needs and desires. Chimps, in contrast, can follow the gaze of another to find hidden food, but they are unable to use intentionally produced communicative cues to do the same (Tomasello, Call, and Gluckman 1997). Chimps fail to understand that ostensive communication is an attempt to attract their attention. They cannot use ostensively and intentionally signaled information from another who attempts to help them by informatively directing their attention in some way (Povinelli 2004; Povinelli and Eddy 1996).

5.5 Language Acquisition in Neonates and Young Children

The infant at 8 to 20 weeks begins to coo when awake and alert (M. Bateson 1979, 1975); the mother responds with soft tones, and turn taking begins;

and by 20 weeks, the infant strings together her coos, but without rhythm or intonation. By 20 to 30 weeks, the infant introduces consonant and vowel sequences, which are repeated frequently with some variation in pitch. Mother responds energetically to these by imitating them and interjecting pauses between individual bursts. This gives the interaction a semblance of dialogue, which both partners find rewarding. Very little simultaneous vocalizing occurs (Bloom and Tinker 2001). Mother distinguishes clearly between vowel-like sounds and more advanced syllabic sounds and cooing or grizzling, which she interprets as signs of comfort or discomfort. The more advanced syllabic sounds are not interpreted as signs of anything but are imitated in interactive, dialogue-type sequences (Papousek and Papousek 1987).

Trevarthen (1979) places the origins of this specialization even earlier, by drawing attention to the small movements of the lips and tongue that he terms "prespeech." Like Papousek and Papousek, he is at pains to point out that such movements occur at different times and appear to be manifestations of a different kind of state and are certainly interpreted in a different way by the mother compared to signs of emotion or physiological need. Language as a vehicle for content appears to have a grounding in species-typical activity patterns that are entirely separate from the vocal-gestural expression of emotion. These species-typical activity patterns are exercised in an interactive context and incorporated into preverbal "dialogues" that serve to select and shape them further (Hendriks-Jansen 1996, 286–287). That language as a vehicle for content is grounded in species-typical activity patterns entirely separate from those of emotional expression is attested by the fact that infants emote *or* talk (Bloom 1991; Bloom and Tinker 2001), up to the emergence of the verbal spurt at around two years (see sec. 5.6).

At about 25 to 50 weeks, the babbling stage commences. Babbling involves the production of sound just for its own sake (M. M. Lewis 2013, 93). Before babbling proper, neonates engage first in cries of discomfort and vegetative sounds (0–8 weeks), cooing and laughter (comfort sounds) typically in response to another's smiling or talking (8–16 weeks), followed by marginal babbling at 25 to 50 weeks (Vihman 1996), where consonant and vowel combinations are repeated with increasing frequency. By 40 to 50 weeks, the babbling becomes variegated, where consonants and vowels are changed from one syllable to the next. The rhythm begins to resemble human adult speech, which parents interpret as speech acts, but the babbling has no obvious meaning. Parents institute games based on melodic sequences and particular syllables, producing protowords associated with specific interactive situations. Children in particular language communities at this age sound more like adults than each other. Babbling consists in small sequences of sounds that are very similar to those used in early language; this implies that the capacity to use the words of native language emerges long before children learn to utter words proper.

These developments in sound production are accompanied or even preceded (often by up to two months) by infant advances in speech perception (Harris 2004; Butterworth and Grover 1989). Infants can make auditory discriminations at a very young age, and some of these are clearly speech related. Infants as young as one to four months can perceive most of the phonetic distinctions that exist in human languages (Miller and Jusczyk 1989). As Hendriks-Jansen points out (1996, 287), these perceptive capacities require subvenient mechanisms of great complexity, since there is no one-to-one relation between acoustic and phonetic segments: "Speakers do not produce speech in well-defined phonetic atoms; the vocal gestures for a given segment will typically overlap those for the segments that precede and succeed it." Human communication, therefore, requires underlying mechanisms of great flexibility:

> We have inherited certain signals like smiling and eye-brow raising from our primate ancestors, but these become assimilated and elaborated into far more complex patterns of social interaction. Early communicative exchanges with adults provide the means for launching an infant into such flexible patterns of "dialogue." Human infants have to be boot-strapped into the world of meaning. Turn taking of the particular human variety is the mechanism that makes it possible. (Hendriks-Jansen 1996, 267)

Turn taking, in this account, can be construed as "the cradle of meaning" (Hendriks-Jansen 1996, 277), but contra Hendriks-Jansen, only because it provides essential opportunities for learning.

Researchers generally acknowledge (Levinson 2006) that neonate and mother are preadapted for speech perception and production. From birth (or even prenatally) neonates appear to be attracted by their mother's voice (Kisilevsky et al. 2009), which is preadapted to the infant's perceptual and production capabilities. Mother alters her pitch, syntax, and pronunciation, resulting in a soothing but highly modulated and repetitive, almost ritualized form of speech. Mother closely monitors the infant's state of arousal (see chap. 4) and adjusts her speech accordingly. She facilitates the infant's perception of her vocal utterances by her timing, that is, by her turn taking in vocalization-pause interactions (see sec. 5.3).

By just two to four months, infants begin to respond differentially to different tones of voice (M. M. Lewis 2013) and, at around six months, associate certain simple utterances with specific situations. Mother, again, adapts her vocalizations to her baby's increasing perceptual and motor abilities; her baby's focus on external objects elicits longer and more elaborate commentaries. In addition, mother begins to ignore some of her infant's vowel-like utterances and attend more closely to her syllabic babbling.

By seven to ten months, infants exhibit a clear preference for "motherese," that is, the exaggeratedly succinct maternal utterances where pauses occur at clause boundaries, and in response, caregivers now exaggerate prosodic changes that naturally occur at clause boundaries. These changes include extended pauses, lengthening of the final syllable, rise or fall in pitch, and increase in volume. Clauses, then, tend to be perceived as perceptual units, allowing them to serve as scaffolding (Scaffolding 1) for later grammatical parsing. What clause perception enables is the ignoring, on the part of the infant, of false starts, "oohs" and "ums," and ungrammatical strings that do not sound like clauses (Hendriks-Jansen 1996, 288). These anomalies are typically identified as examples of the difficulties children must experience in abstracting grammatical structure from defective strings, and therefore as providing grounds for the rich innate grammatical structure hypothesis.

What this implies is that speech perception and production are enabled by a range of prelinguistic activity patterns that have specifically been adapted to serve as a potential vehicle for content. From the beginning, and up until the vocal spurt, at least, the vocal behavior that will eventually be appropriated for the production of contentful sentences is clearly differentiated from the vocal expression of emotional and physiological states. A species-typical capacity to produce a variety of language-like sounds, which matures gradually during the first year, supplies the raw material from which the mother can shape primitive dialogues in much the same way as precursor emotions provide the raw material for her to shape her baby's basic emotions and, eventually, her HCEs. These species-typical activity patterns provide the means for initiating and sustaining vocal interactions with the mother, which serve as scaffolding (Scaffolding 2) within which the infant can practice the basics of speech acts.

It is interesting to note that, before the emergence of the vocal spurt, infant vocalizations are not tied to emotional stimuli. Language learning is a game. Mothers attempt to engage their young ones in dialogue only when they are not over- or underaroused (Bloom 1991). Language is learned through interactive play where emotions may be represented symbolically, but not through the attempt to express strong emotions. This clear separation between signals of immediate need and the interactive emergence of discursive language will eventually permit language to be about things that are not present and even counterfactual (Hendriks-Jansen 1996, 289).

In sum thus far: Before the infant learns to comprehend and produce her first words, she has become surprisingly skilled at conducting conversations. She is able to distinguish questions, statements, and commands in adult strings and produce them herself by varying the pitch, volume, and rhythm of her vocalizations. She has become skilled in turn taking, having learned to imitate and respond to adults' phrasing by using vocal and nonvocal cues. As a result, the infant can follow the conversation of adults although she does not fully comprehend the explicit meaning of what is said. She has learned a range of ritualized games involving mother, objects, and herself. Finally, toward the end of her first year, she will have learned some words that stand for specific interactional contexts and games.

This prelinguistic dialogic interaction is what Spurrett and Cowley (2010) and Cowley (1998) term "utterance activity," and it includes words and strings of words but is not restricted to them. It is also viewed as "continuous with and inextricable from (nonwritten) language" (Spurrett and Cowley 2010, 296) and persists throughout the life span. Cowley claims that it challenges the hypothesis that language is only a system of turn taking, and as such, his views are consistent with those of Fogel on coaction (see chap. 4). Cowley provides an interesting empirical example: The mother, father, and daughter of an Italian family are engaged in an "utterance activity" that results in the almost simultaneous pitching and timing of an exclamation while gently and good-humoredly mocking the mother. The exclamation is "Oeu!" which means something akin to "Come on!" in English. Father's "Oeu!" begins while Mother is midstream and ends pitched at about one octave above his normal to accurately match hers. Before

Father stops his "Oeu!" his daughter joins in with an "Oeu!" of her own, extending longer than his but dropping her pitch to the top range of his before she stops speaking. She then gives a short laugh ("Ha!") at that pitch. In this interaction, the utterances of father and daughter are in harmony, in both pitch and timing, and together they gently mock mother without any need for a single standard "word" (Cowley 1998). Accent, timing, loudness, and various kinds of visible movement, along with a host of other pragmatic influences, allow adults to regulate each other's speech. This type of influence, it is claimed, is ubiquitous in utterance activity at all ages and occurs in word-based speech as well (Spurrett and Cowley 2010; Cowley 1998).

If this account is correct, then utterance activity is metaphysically widely realized. The utterance activity involves mutual and synchronous modulation, the locus of control shifting from the interlocutor who initiates the utterance activity, to shared "environmental" control, then shifting again to the interlocutor who terminates the interaction. I return to this phenomenon later, when discussing the role of saccadic vision in utterance production and comprehension.

By imitating parents, infants learn interactive timing, clause boundary recognition, and the rhythms, stresses, and intonations associated with commands, requests, statements, and questions, which are extensively practiced before the child learns her first words. These are all ontogenetic predecessors of fully linguistic communication in that the basic structure of human communication is already in place before toddlers begin to speak intentionally (Wyman and Tomasello 2007, 472). What sustains and shapes this repetitive activity is the interactive context of mother-infant dialogue, which is experienced as mutually highly rewarding.

At about one year, children start to use one-word language forms that are related to the people closest to them (*Mama*) and to routine daily activities (*din-dins, hello, bye*); that is, they refer to people, objects, and events that are perceived in mundane dynamic relationships (Bloom 1991). What this and the detailed account of the development of conversational pragmatics given earlier indicate is that the linguistic environment into which neonates are born and in which they develop scaffolds the development of children's language and conceptual thought in much the same way as the physical environment scaffolds the development of motor capacities. From the one-year-old's learning of simple labels, such as "Dada," to the narrative soliloquies of four-year-olds, language allows the articulation of objects, people, and events perceived and the progressive refinement and elaboration of such articulated percepts.

When the child first learns to use words, the nature of the intended speech act is conveyed by prosody and gesture. "Dada" spoken with rising intonation means "Where's Daddy?"; "Dada" spoken with falling intonation means "That's Daddy"; and "Dada" uttered with insistent monotone means "Daddy, please give me some attention." It is clear that much of the meaning of these utterances depends on the conversational context in which they are uttered and the interpretive skills of addressee caregivers. Even when the child progresses to two-word forms at around eighteen months, context remains critically important (as it does for most communication even in adult life; see sec. 5.6 and chap.7).

One- and two-word utterances can act as vehicles for content because the child and her parents have shared experiences of the situations, events, and activities that routinely and regularly take place within them (Hanks 2006). One-word sentences, however, restrict the speaker's communicative abilities to the immediate context. "Single word commands and requests will not get you very far if you want to convey things *about* a routine or situation. In order to move beyond this stage in his speech, the child must learn to detach the linguistic object from the tight embrace of its grounding, so that it may be used combinatorily as an atom of meaning, with abstract markers for the relations into which it can enter. Language itself provides the scaffolding for this next stage of development" (Hendriks-Jansen 1996, 293; his italics).

Around eighteen months, the child begins to use two-word utterances and has begun to attend to the internal structure of her own and other people's utterances. The existence of a public language with a grammatical structure allows the child to produce indicative and imperative icons as opposed to performing vocal gestures as part of the interactive process. "Teddy cry," "There Daddy," and "Kick ball" all mean something that is not tied to the immediate needs and interests of the child. In addition, children at this age provide a running commentary (learned by imitating mother) on any new or unusual activity (Eckerman 1993). They make sense because adults can use the structure of natural language, as well as the context of utterance, to disambiguate meaning, allowing them to respond to the child's utterances as declarative statements rather than as signs of internal homeostatic status.

I have been at pains in this (and the preceding) chapter to emphasize the "tightness" of the infant's world; this tightness, I have argued, is a function of inbuilt perceptuo-motor constraints and inbuilt caregiver tendencies to scaffold or bootstrap development. This mechanism is thought by some (e.g., Bruner 1976) to include concept formation, which begins in repeated, routine interactive activities and continues with linguistic bootstrapping from caregivers. Bruner (1976, 200) describes the beginnings thus:

> Much of the child's early mastery is achieved in oft-repeated tasks. The child spends most of his time doing a very limited number of things. There is endless time spent in reaching and taking and banging and looking, etc. But within any one of these restricted domains, there is a surprising amount of "systematicity." It consists of two forms of "playful" activity: in one, a single act (like banging) is applied to a wide range of objects. Everything on which he can get his hands is banged. In the second type, the child directs to a single object all the motor routines of which he is capable—he takes hold of the object, bangs it, throws it on the floor, puts it into his mouth, puts it on top of his head, runs it through his entire repertoire.

Such activities are unified merely by spatial and temporal contiguity (although this is not Bruner's view), themselves based in the limited motor repertoire of infants and the limited range of toys that are made available to them. Such toys will be characterized by their bright colors and interactive potential (they will move or sound in interesting ways when variously manipulated). As Hendriks-Jansen (1996, 292) observes, the environment of young infants is constructed to facilitate, that is, to scaffold the activities of grasping, banging, and throwing.

True concepts, however, that can be separated from origin and combined with others require linguistic scaffolding. As discussed earlier, the child at eighteen months can already engage in "conversation." She has learned that certain sounds are reliably associated with certain objects, events, and situations through continual exposure to baby talk commentary and, subsequently, to commentary tailored to her comprehension capacities. Deliberate linguistic scaffolding by parents who diligently correct pronunciation (to be imitated) results, at about one to two years, in infants' learning about ten new words a week; from two to ten years, however, during what is considered to be the vocal spurt, this increases to ten new words a day (Feldman 2006). By two to four years, children justify deviations from the norm in terms of narratives involving intentional ascriptions, and they spend appreciable periods of time soliloquizing about their own and other people's activities. They are not simply reporting these activities (K. Nelson 1989); they are trying to make sense of the way they themselves and other people have acted. This obviously includes emotion-related narratives pertaining to self and others (see chap. 4). Making sense of one's own emotions, actions, their causes and feelings, and those of others, that is, making sense of what is relevant to her, provides the child with a platform on which to build her intrapersonal control. All of this is entirely consistent with a relevance-theoretic approach to human communication, which I discuss in chapter 7. Of central importance to that approach is the use of "loose talk" by way of extension broadening or narrowing. Loose talk, however, is not something that is useful only to sophisticated language users; rather, it commences in early childhood.

Beginning at two to three years, children show an interesting creativity in their use of language. If the child does not possess a word in her vocabulary to convey what she wants to convey, she will use a word that she does possess that is reliably associated with experiences that are similar, such as objects of the same shape or animacy (McDonough 2002). This is called "overextension." It is, in an important sense, the beginning of loose talk, where children produce ad hoc meanings to provide evidence of their intended meaning. They can understand and produce metaphor once their conceptual and linguistic capacities (and the experiential world in which they interact requires them) are sufficiently developed and sophisticated to produce and understand complex implicature.

5.6 The Eyes Have It

> A proper understanding of how language processing works requires attention to the perceptual motor patterns that interface the organism with its environment (Zwann and Kaschak 2007) and those environmental properties in which the language use is embedded. Without the many different environmental constraints that situate language processing, too often lumped together under the monolithic term *context*—language as we know it would not exist.
>
> —Spivey and Richardson (2009, 384)

Spivey and Richardson (2009) report a range of eye-tracking experiments that demonstrate that the visual world is continuously accessed and integrated with spoken language processing, at a number of levels. This is one particular example of saccadic vision's more general function of exploiting external memory (world). Eye tracking is involved in word recognition, reference resolution, syntactic processing, and thematic role assignment. They argue the following points:

(i) Even during the few hundred milliseconds it takes to hear a spoken word, situational constraints, that is, objects visually present, can influence the activations of lexical representations that result from hearing the first few phonemes. In an experiment that included both candy and candle in the visual display, subjects fixated on an object that shared the same first phonemes with the target object, that is, candy and candle. When the same subjects were shown a visual display with only a candy and no candle, they fixated on the candy and rarely on other objects. Spivey and Richardson also cite studies where the actionable situational constraints determined comprehension. When subjects were asked to put a cube into a can, and two cans were available, one large enough to accommodate the cube and the other too small, the size of the cans immediately constrained the referential domain of the can. Subjects looked only at the appropriately actionable object despite the obvious ambiguity of the utterance. These are just two examples of the several that Spivey and Richardson cite to demonstrate clearly that the external world is visually interrogated continuously throughout the course of continuous language input. The two I mention here relate to word recognition and reference determination, but Spivey and Richardson also include interesting experiments relating to syntax parsing and thematic role assignment (which are superfluous to the needs of this discussion).

(ii) A range of other situational factors influence language comprehension. Knowledge of the speaker's occupation (e.g., which sense of "case" is appropriate when uttered by a judge or porter), the speaker's identity (e.g., gender, ethnicity, social class), and the speaker's actual visual field are all instrumental in utterance comprehension. Situational factors also include assumptions about the speaker's past experience and beliefs. H. Clark (1996) terms these assumptions "the common ground," and it is, of course, analogous to Sperber and Wilson's cognitive environment (which I discuss in chap. 7). In addition, interlocutors actively manipulate the common ground to facilitate communication by, for instance,

> pointing and gesturing, both of which are visually processed. An additional and important fact is that the degree to which conversants can coordinate their visual attention moment to moment across a shared visual display is causally related to the success with which they communicate. This is a useful example of the operation of joint attention, which develops very early in ontogeny. (This is implied, I think, by Sperber and Wilson, but they do not specify coordinated visual attention in the cognitive environment.)

Spivey and Richardson conclude: "As part of the linguistic process of recognising words or parsing syntax, participants make saccades to referents in the world that are phonetically, semantically or pragmatically appropriate. In conversation, the degree to which a listener follows a speaker's gaze around the world is an indication of their understanding, and in face-to-face conversation, eye movements can serve as linguistic cues" (2009, 296). This is because visual object fixation by the speaker referentially fixates the utterance for the listener; that is, visual object fixation (or display interrogation) ostensively evidences what is relevant to the listener in the utterance. These nonlinguistic ostensive skills are learned very early in ontogeny and provide the scaffolding that enables the acquisition of natural language.

Two more points relating to the context dependence of utterance comprehension are worth noting. The first is that in many East Asian languages, it is sometimes acceptable to omit any elements that can be determined, that is, inferred, from context. Japanese children, for example, are explicitly taught how to determine meaning from context as part of their language education. In addition, adult Hebrew is written without vowels, but readers effortlessly pick (i.e., infer) the right words (Feldman 2006, 265).

5.7 Neurochemical Underpinnings of Human Prosociality

Researchers have suggested (Tomasello and Carpenter 2005) that human biological adaptations result in (i) an understanding of intentional action and perception, which coevolved in chimps in the context of intraspecific competition, and (ii) skills and motivation to share their psychological states with others. These skills and motivations are unique in human beings and evolved in the context of intensely cooperative activities of a particular kind (those that increased hominid fitness). Prosocial skills and motivations depend on appropriate circuit neurogenesis, maturation, and connection, which, in turn, depend on intraneural genetic activity, and IEGs were identified as one type of gene involved. This genetic activity is excited by what Panksepp (2007) terms prosocial neuropeptides and endogenous opioids.

Intraneural genetic activity is excited by endogenous opioids, which themselves are potentiated in their activity, particularly in terms of duration, by certain social neuropeptides and more generally active, appetitive neurochemicals (E. Nelson and Panksepp 1998). The former group includes oxytocin, vasopressin, and prolactin, and the latter serotonin, dopamine, and norepinephrine. Opioids are released during all prosocial activities, such as being held, being groomed, and playing, and, as opioids, induce feelings of great pleasure (Panksepp 2007). Oxytocin has been found to be particularly active in terms of producing maternal "urges" and in the social bonding of neonates and mothers; it has also been shown to generate feelings of trust and love (Panksepp 2007; Zac, Matzner, and Kurzban 2008) and lower social fear (Kirsch et al. 2005). It is also released during suckling (with prolactin) and at other times (e.g., orgasm, parturition).

This neurochemistry explains the mutual pleasure both infants and mothers garner from their various interactions, how the very close relationship (bond) develops between them, and why social interaction is enjoyed simply for its own sake. It also explains why caregivers feel compelled to respond to the unfocused crying and motor unrest of their distressed neonates. This neurochemistry activates neuron producing, maturing, and connecting genetic activity that provides the neural basis for more advanced socioemotional and cognitive activity.

Some interesting implications arise from the foregoing; I begin with the most obvious:

(i) The development of full human emotionality takes places simultaneously with the development of intentional thought and language and through precisely the same mechanisms, that is, through the progressive epigenetic Scaffolding (2) of mother-infant interaction, the initial establishment of which is enabled by inbuilt sensorimotor constraints and species-typical activity patterns. This scaffolding provides the necessary and sufficient conditions for the progressive development of the infant's emotion systems, including her emotional thoughts, nonemotional thoughts, and their linguistic expression. Full human emotionality, I have argued, requires the semantic conceptualization of emotion, and segmented language enables both the prerequisite concept formation and the linguistic expression of emotion. Ever-increasing complexity in hominid social life provided ever-increasing complexity in the number and nature of potential emotional elicitors, and these selected for ever-more-sophisticated emotion and emotion-expressive systems. It is entirely reasonable, therefore, that emotional complexity evolved with intentional complexity and the means to their expression, that is, segmented language. The acquisition of language is explicitly pedagogical and is timed to avoid emotionally charged situations, both of which facilitate the learning of words. It seems unremarkable, therefore, that the linguistic expression of emotion begins only with the word spurt; prespurt children simply do not possess the words with which to express emotion (and their emotion expression signs can manage without them at this stage of development).

(ii) The key components of the simultaneous development of full human emotionality, intentionality, and language are species-typical activity patterns and cultural scaffolding. The findings from research in developmental psychology canvassed earlier identify a number of activity patterns that are unique to human beings and seem to have evolved to elicit the attention of caregivers. This is suggested by the suckle-pause-jiggle suckling patterns described earlier. The maternal jiggling response, it was noted, is not grounded in any biological need; the infant normally resumes suckling spontaneously after 15 to 20 seconds. This suggests that what was selected for is its pivotal function in establishing turn taking, on which the psycho-emotional and perceptuo-motor development of the infant will crucially depend. In addition, these species-typical activity patterns are driven by primitive intentional devices, and not any full-blown representation of the world. They are biological preadaptations to which caregivers are preadapted to respond.

(iii) The suckle-pause-jiggle species-typical interactivity pattern is clearly not selected for nutritional purposes; Hendriks-Jansen (1996, 266) suggests (reasonably, in my view) that its selection had to do more with establishing the pragmatics of communication than with feeding. In addition, it is the caregiver's intentional interpretation of the pause that enables the jiggle interaction to play the scaffolding role that it does. It provides the basis for launching the infant into an intentional, linguistically mediated adult social world. I have argued, however, that turn taking provides opportunities for motor mimicry and feedback regarding emotional signs, motor skills, and language. It is the learning opportunities that turn taking provides, not turn taking per se, that launch infants and children into the intentional and linguistically mediated adult world. This is consistent with turn taking shading into coregulation as mother and baby repeatedly interact in a limited range of activities.

(iv) It is clearly efficacious for neonates and infants to possess primitive and limited forms of behavior in early development. It has already been noted that the infant's visual limitations in both object resolution and fixation (together with the reciprocal maternal preadaptation to position her face at the appropriate 20–24 cm distance) result in an experiential world composed almost exclusively of the maternal face. This enables the establishment of face-to-face interaction on which so much will depend developmentally. Similarly, the infant's inability to distinguish separate words in her mother's vocalizations may allow the infant to respond to her clauses as unitary utterances, equivalent to her own coos and murmurs, and facilitate the process of turn taking in early dialogues. Both of these limitations provide the basis for the later scaffolding of parsing of clauses into meaningful units (M. Bateson 1979). They enable the infant to ignore "ums," "ahs," and false starts and any ungrammatical strings that do not sound like clauses. This sort of anomaly is considered sufficient to preclude the acquisition of correct grammatical structure from imperfect strings and led to the positing of a rich, innate language acquisition device (LAD). Yet these limitations are actually advantageous in the scaffolding of language acquisition. The infant's inability at this stage to attend to the fine structure of maternal utterances allows her to attend to the gross features of the utterance, that is, its intonational and rhythmic features. A premature ability to distinguish syntactic and semantic detail would render the acquisition of these pragmatic features very difficult, and it is these intonational and rhythmic features on which syntactic and semantic speech acts are built. In addition, the child's brain would be overwhelmed with detail and incapable of sustaining the interactive context out of which her more sophisticated language abilities will emerge (M. Bateson 1979, 68).

(v) A great deal of the cultural scaffolding provided by such interaction is clearly of the Scaffolding 2 type. I noted in the previous chapter that infant-maternal interaction is best characterized as coactions where it is impossible to identify which partner is controlling or leading the interaction; mutual gazing was the example offered here. Fogel (1993) provides other examples, including object passing and smiling. When a child attempts to hand an object to her mother, it is widely assumed that the child is signaling a desire to the mother, who interprets the behavior as such and extends her hand to receive the object. The infant correctly interprets this signal and drops the object into her hand. Such an analysis is typical of Scaffolding 1 and is underpinned by folk psychological interpretations. The infant desires to pass the object and thus controls the interaction. However, when captured on slow-motion video camera, the interaction does not support such an interpretation. Mother and infant are moving simultaneously toward the point of mutual contact, but that point is not any predetermined location; rather, it is dynamically constituted by the continuous coregulated actions of both partners. It is impossible to identify who is sender and who is receiver. This interaction is clearly an example of Scaffolding 2. The same is also true of smiling. The typical folk psychological interpretation of a smile is that it conveys a specific message that originates with the smiler, who adjusts her facial expressions, body movements, and gaze to the nature of the message she wishes to convey (e.g., joyful, encouraging, or merely being polite). Again, the folk psychological interpretation is that the smiler's intention shapes the nature of the smile and determines the message it conveys. Once again, however, videotaping demonstrates that this is not the case. The smile emerges in an interactive context, and its nature may change at any point depending on the reaction of the receiver. Smiler and receiver continuously and synchronously adjust their facial expressions, bodily movements, and gaze in response to the other. The smiler's mental state is relatively unimportant once the smiling interaction is under way (Fogel 1993).

(vi) I have already argued that these species-typical activity patterns are primitive intentional icons, and mothers are correct in interpreting them as such (pace Hendriks-Jansen). These examples, however, should not be taken to suggest that full-blown folk psychological ascriptions are always mistaken; this is clearly not the case. In adult interaction (as the previous analyses of ontogenetic emotional and cognitive development attest), people do ascribe mental states like belief and desire to others. It is this ascription that renders their behavior goal directed and intelligible and makes social intercourse possible at all. What I suggest, rather, is that activity patterns of any sort, including interactivity patterns, require initiating, and this is necessarily a function of intentional icon, even if only one that is genetically hardwired (like "punctuated" suckling). The full-blown intentional icons of adult human beings, those that are beliefs and desires, can be as instrumental in initiating species-typical activity patterns as any that are hardwired. Individual autonomy or internal locus of control depends on this. These examples are simply illustrative of situated cognition. An agent "decides" to act or interact in a certain way and for a certain purpose, but once she has decided, she can leave it up to her highly trained and flexible subsystems to carry out the decision. Without this fluid situational flexibility, smooth and seamless interaction would not be possible.

(vii) The above analyses demonstrate that human beings require a protracted apprenticeship into sociocultural life to acquire folk psychological explanations for behavior. Their apprenticeship is scaffolded by the intentional interpretations of their mothers, whose own apprenticeship was scaffolded by similar intentional interpretations of their mothers. The progressively more sophisticated (and initially heavily scaffolded) interactions provide the basis for the development of a full-blown intentional folk psychology from the minimally intentional icons with which human infants are endowed. It is by participating in intentional activity of differing levels of complexity and explicitness (i.e., those of neonate and caregiver) that human infants become intentional agents in the fullest sense.

(viii) The total realizer system of human emotional ontogenesis includes a host of heterogeneous intra- and extrapersonal elements (at least two bodies and two brains, deeply integrated into linguistic and artifactual cultural ethnotheories and practices), and the core realizer, obviously, is the special relationship that mother and infant progressively coconstruct. The locus of control in human ontogenesis changes from environmental to shared to individual. Again, however, either partner can trigger interaction once the infant's neural hardware is in place, and either can terminate it, but once online interaction is up and running, the locus of control is shared. The degree of complementarity, complexity, and integration, from inborn constraints to behavioral patterns, to mutual, synchronous modulation, from interpersonal emotional to intrapersonal regulation, is truly remarkable.

5.8 Summary

The acquisition of a pragmatic foundation and semantic lexicon are as dependent on the close, linguistically mediated social relationship of neonate and caregiver as is emotional ontogenesis, with which it is so intimately connected. The preadaptations with which both partners are genetically endowed, that is, neonates' species-typical behavior patterns and parents' intuitive parenting skills, evolved precisely to promote mutual engagement. It is through the coactions of baby talk commentary and speech modeling, in particular, and motor mimicry (including pragmatic features) and mutually rewarding feedback that coarse-grained search-identify-satisfy interpreter devices of caregivers are progressively fine-tuned, and coarse-grained producer devices associated with protospeech and other species-typical behavior patterns are progressively fine-tuned. It is through this interdependent (deeply functionally integrated) process that contingencies between objects and events, pragmatic features, and semantic content (including that relating to emotion) are synchronically constructed in both partners to differing extents. The intensity and frequency of the interaction are sustained by the extremely pleasant feelings generated by the release of endogenous opioids and prosocial neuropeptides, which also excite intraneural genetic products into neurogenesis and neural connection.

The level of deep functional integration of both neonatal and caregiver neural and corporeal resources with external sociocultural resources in language development is just as deep as that of emotional ontogenesis; indeed, many of the resources are influential in both types of development. The level of deep functional integration is such that they constitute a genuinely extended cognitive system by affecting the system synchronously as a whole and producing effects that result only from the system functioning synchronously as a whole. This demonstrates, again, that it is a mistake to construe causality and constitution as mutually exclusive relations; the causal relations underpinning the development of language, as well as that of emotionality, are clearly constitutive of larger, more complex structures and processes. In addition, given that the same processes are involved in language development and emotional development, language development is also a world-to-brain transcranialist achievement.

6 From Evolution to Emotionese

6.1 Introduction

In chapters 3 and 4, chapters focusing on the metaphysics of emotion and emotional ontogenesis, respectively, I argued that basic emotions and higher cognitive emotions both develop from innate emotional precursors in the context of a close, linguistically mediated social relationship, and that they are widely metaphysically realized, the realization of emotional ontogenesis being radically wide. In those chapters, moreover, I argued that human emotional ontogenesis proceeds from the triggering of innate proprioceptive and interoceptive devices by positive and negative absolute stimulus thresholds, through interpersonal emotional regulation with caregivers, to childhood intrapersonal regulation. I also argued that innate caregiver capacities of exaggerated and succinct affect mirroring, augmented by baby talk commentary and informed by relevant ethnotheories, together with neonates' and children's sensorimotor preadaptations and mimicry, were the main vehicles for such development. Such innate complementary endowments enable the memetic transmission of parental emotional and more general ethnotheories to the next generation to regulate the social interaction of individuals and groups. Emotional ethnotheories ensure the appropriate observance of, and compliance with, the more general ethnotheories of the culture. Similarly, in chapter 5, I argued that a range of species-typical behavior patterns interacted with intuitive parenting skills, most of which were also operational in emotional ontogenesis, in the development of language. The interaction of these complementary repertoires results in the development of the close relationship that provides the context for further development.

I emphasized in these previous analyses that both the emotion precursors and the species-typical behavior patterns of human neonates are the results of brute physiological processes that are interpreted as or function as signs of psychophysical or homeostatic status by preadapted caregivers. That this is universally the case in all intact, normally functioning human caregivers points to a selectionist history: emotion precursors and species-typical behavior patterns are produced by devices that were selected for in ancestral environments because they conferred a fitness advantage on their bearers. In short, emotion precursors and species-typical behavior patterns are signs or representations of psychophysiological states that are produced by devices whose purpose or function is to produce them. Similarly, the intuitive parenting skills that are triggered in caregivers by precursor emotions and species-typical behavior patterns are produced by devices whose purpose or function is to produce them. Accordingly, one of the foci of this chapter is function; the theory I adopt here is that of Millikan (1984), which is consistent with previous analyses of emotional and language development and metaphysical realization. Of particular interest to human emotionality are Millikan's theories of relational, derived, and adapted functions.[1]

In addition, and since emotion precursors and the discrete emotions proper are both natural signs and intentional signs and, when expressed linguistically, linguistic signs, my second focus is the biosemantics of mental content. I discuss the distinction between these different sign types, and how one emerges from another. I argue that natural signs carry local natural information relevant to individual (and group) well-being, "local" here referring to a natural reference class (or domain of reference), and when intentional and linguistic signs are functioning properly, they do, too. These analyses are largely informed by Millikan's biosemantic theory. My indebtedness to Millikan is obvious; however, the chapter is not merely exegetical. My analyses are original in that they are the first, to my knowledge, to apply Millikan's theories to the development of human emotionality and language.

In chapters 4 and 5, I was concerned to show, in particular, the deep functional integration of sociocultural caregiver and neonatal intracranial resources. The content of this chapter does this, too, but it demonstrates with additional precision the utterly breathtaking complexity of the functional integration of these resources.

Emotions and language, I have repeatedly claimed, develop together and largely through the same mechanisms. In light of this, and to keep the analysis manageable, I restrict my focus to emotional development. What I say about emotional development, therefore, will also be true of language development.

I conclude the chapter by rebutting the intracranialist claim that transcranialists fail to recognize the mark of the cognitive. Transcranialists do not fail to recognize this putative mark. Most recognize a certain kind of information processing as being the mark of the cognitive; others of a teleological or biosemantical inclination, such as myself, while agreeing that information processing is what cognition essentially does, also deny that there is anything in the head that is intentional just as such. I begin my analysis, however, with some rock-bottom preliminaries with respect to theories of function.

6.2 Theories of Function: Rock-Bottom Preliminaries

According to *The New Shorter Oxford English Dictionary* (1993, 1042), a function (derived from the past participle of *fungi*, "to perform") is "the activity proper or natural to a person or thing: an office, duty, employment or calling. Also, a particular activity or operation (among several); an organisational unit performing this." This (these) definition(s) subsume two distinct concepts of function extant in the philosophy of science and biology, namely, what have come to be known as proper (evolutionary, biofunctions, or teleofunctions) and Cummins functions. The term *proper function*, from the Latin *proprius*, meaning "one's own" (Millikan 2002, 116), is due to Millikan (1984, 1989a, 1989b, 1993, 2002, 2004, 2009); and *Cummins function* is the term given to function in Cummins's (1975) causal functional analysis.

These two concepts have traditionally been considered to operate in distinctly different explanatory projects; proper functions operate in evolutionary theory and behavioral ecology to explain the presence of a functional trait in its bearers (and, because of this, are classified as etiological theories), and Cummins functions operate in physiology and developmental biology to explain the causal contribution of a functional item to a complex process (Millikan 1989a; Godfrey-Smith 1994) and, because of this, are classified as dispositional theories. Some scholars (e.g., Buller

1999b; Price 2001; Millikan 2002) now believe that, following a range of suggested amendments, the two types of functions do not operate in distinctly different explanatory projects but operate in a complementary fashion in all explanatory projects concerning biological items. The views of Millikan on reconciling proper and Cummins functions include complex discussions concerning exaptation and are well beyond the requirements of the present discussion. All that is required for the present discussion is a minor tightening of the theory of proper function by adding Schwartz's (2002) continuing usefulness requirement.

Etiological theories of function argue that the functions of biological devices are grounded in their evolutionary histories. The function of an intentional device depends on the function of the mechanism that produced it. In addition, the function of an intentional mechanism is to guide the operation of some further mechanism; for instance, an executive mechanism will control some aspect of the organism's behavior in such a way that the behavior of the second mechanism is coincident with some favorable condition in the environment. What this immediately implies is that because it must fit or respond to the producer mechanism, the second, cooperating device is partly determinative of the producer device in terms of function and content. The tongue-snapping mechanism of the northern leopard frog, *Rana pipiens*, is instructive: the mechanism exists to ensure that the tongue snaps when a fly is within range of the tongue snaps. To snap at flies when they are within range is the proper function or purpose of the tongue-snapping mechanism (Price 2001; Schwartz 2002).

Two philosophical problems are associated with biological teleology. The first relates to the concept of function itself, and the second to efficient causation (Buller 1999a). Biologists routinely refer to the concept of function in their descriptions of organs and traits of organisms; for example, the function of the heart is to pump blood, the function of the kidneys is to filter metabolic waste from blood, and so on. Such function statements cite effects of organs or traits, but not every effect of a trait or organ corresponds to a function of it. The heart makes lub-dub noises, now useful in cardiothoracic diagnostics, but making lub-dub noises is not the function of hearts (biologists agree that the function of the heart is to pump blood). This implies that "the function of X is to Y" does not simply mean "X produces the function Y," because it would fail to distinguish the effects that it is the function of the trait or organ to produce from those it is not its function to produce. What this further implies is that function does not appear to be wholly explicable in terms of ordinary causation. Buller (1999a, 2–3) suggests that Hull (1974, 102) puts the matter "very nicely":

> Just as a physicist might say that heating a gas causes it to expand, a biologist might say that heating a mammal causes it to sweat. But a biologist might also say that a mammal sweats when heated in order to keep its temperature constant, while no physicist would say that a gas expands when heated in order to keep its temperature constant, even though that is exactly what happens.

The question, then, is this: why is the constant temperature merely an effect (i.e., caused by) the expansion of gas, whereas it is the or a function of sweating in mammals? Explicating the biologist's concept of function is required to answer these questions.

In terms of efficient causation, it is plausible to construe the distinction between sweating mammals and expanding gases in terms of *purpose*; mammals sweat *to bring about* body temperature constancy, but gases do not expand for the purpose of maintaining constant temperature. Sweating in mammals, therefore, is a teleological process, whereas expansion of gases is not. The question, then, is this: how can a process like sweating be purposive, or goal directed? The metaphysical problem of teleology, therefore, is to explain whether, and if so, how, there can be purposive, goal-directed processes in a universe governed entirely by efficient causation. How can the biological concept of function, which is "prima facie infected with final causation" (Buller 1999a, 6), be rendered compatible with an exclusive, efficient-causation worldview?

Millikan deals with these two issues, and I show how she does so below. I discuss her theory and highlight its strengths and limitations. In light of the challenges to Millikan's "pure" theory, however, I include Schwartz's (2002) continuing usefulness requirement, which Millikan herself endorses (Millikan 2002).

6.3 Millikan's Proper Functions

In a nutshell, proper functions are functions of reproduced or copied items; they are functions that were selected for by evolutionary processes because they increased organismal fitness. This implies that (i) an item's proper function is what it does on purpose, rather than accidentally; (ii) the term *proper function* is normative in a nonevaluative way; an item retains its proper function even when it fails to perform that function, temporarily or permanently, from malfunction, defect, or lack of enabling conditions (for example, a sperm requires an ovum to perform its proper function, and a diseased or defective heart is still a heart); and (iii) the proper function of a device or system explains its existence, that is, it explains why tokens of certain traits or properties exist at all.

The details of her theory are rather more complex. According to Millikan, an item has a function only as a member of a reproductively established family, which she distinguishes into first-order and higher-order reproductively established families. First-order families consist of replicators, which directly reproduce their structures when making copies of themselves in accordance with the causal laws of nature. This means that each temporal sequence of replicators that are related by descent (through copying) forms a first-order reproductively established family. The copies of the immediate early genes for emotional neurogenesis, for instance, are members of a first-order reproductive family. Replicators do not only cause their own replication, however; they also cause the construction of the vehicles that contain them. As vehicles reproduce, therefore, not only are the replicators they contain reproduced, but the properties of the vehicles built by those replicators are reproduced, too. In short, replicators are directly reproduced through copying, whereas the properties of vehicles are indirectly reproduced via the reproduction of the replicators that construct the vehicular properties (Buller 1999b). Such indirectly reproduced properties of vehicles form higher-order reproductively established families, examples of which include traits such as certain emotional neural substrata, blood type, and body organs (livers, hearts, etc.). Members of a higher-order reproductively established family are thus related by descent also, via the direct relations of descent of the members of the first-order reproductively established family that produce the members of the higher-order family. (This presumably accounts for the term *higher order,* which Millikan fails to explicate in detail.) In both cases, therefore, the *ancestors* of a particular member of a reproductively established family are those temporally prior family members to which it is related by a continuous chain of direct or indirect reproduction (Millikan 1984). In short, it is the function of X to do Y if doing Y caused the proliferation of ancestors of X (through either direct or indirect reproduction).

The important point to note here is that a trait or property does not possess a function unless its ancestors were selected for in the lineage's evolutionary history. It is the function of a member X of a reproductively established family to do Y if ancestors of X did Y and doing Y causally contributed to their family's having greater reproductive success than competing reproductively established families and, hence, causally contributed (eventually) to the production of Y. What this clearly implies is that trait T produces function F only when (a) T is hereditary, (b) there has been variation in T within a common selective environment, and (c) the bearers of T had a greater fitness within that common environment than bearers of T's variants (at least partly) because of possessing T (Buller 1999b). These are all consequences of a strong adaptationist view that all beneficial effects of all traits of all proper functions are selected for, and selection always operates on variation. Godfrey-Smith (1994, 356) rightly points out that this imposes a "substantial biological commitment" that the account cannot support.

The substantial biological commitment that the strong theory cannot support is due to four serious problems: (i) despite the claim that only beneficial effects can be selected for, at least one type of gene, the segregation distorter gene, disrupts cellular meiosis, and this explains its survival (Godfrey-Smith 1994, 348); (ii) it eliminates the functions of hereditary traits that are produced by genetic drift yet still contribute to their bearers' fitness (Brandon 1990); (iii) it eliminates the functions of partial realizers that remain invariant in a total realizer system; this could occur when a complex trait T is constituted by ti, tii, ... tn where T is selected for but where any one of ti to tn remained invariant (Buller 1999b, 290); and (iv) it fails to account for changes in function (Godfrey-Smith 1994) and loss of function as in vestigial traits (Griffiths 1992, 1993). These flaws are addressed by Schwartz's (2002) continuing usefulness account, to be discussed subsequently.

Millikan's theory has two important strengths, however. First, it quite clearly addresses the philosophical problem of teleology. It shows how "the function of X is to do Y" can causally explain the existence of X without invoking backward causation; it is not the fact that X itself produces Y that explains X's existence (backward causation) but the fact that *ancestors* of X produced the effect Y and that Y was among the causes of the existence of X (via the process of replication and development that eventually produced X).

Second, Millikan's theory explains how teleology can emerge within a universe governed solely by efficient causation. In the beginning of the biological world, the only vehicles were probably replicators differentially replicating themselves in primeval soup, where their differential replication was due to differences in what they did in interacting with their environments. The first functions emerged, at this point in the universe, as functions of members of first-order reproductively established families. Eventually, however, some replicator interactions resulted in types of replicators combining together to form more complex vehicles consisting of combinations of replicators. The properties of these complex combinations of vehicles biased interactions in their favor, and as a result, they began to enjoy greater differential success in replication. Progressively, more and more replicators combined, and protein walls were formed for protection, and thus the first cells appeared. Multicellular vehicles eventually followed, and finally the complex organisms that we now take to be paradigmatic vehicles. In addition, the emergence of these higher levels of complexity was accompanied by the emergence of the functions of members of higher-order reproductively established families (Buller 1999a). This provides an explanation relating to the origins of complex organisms but refers to nothing but efficient causation and the differential perpetuation of replicators (which is a function of differential success in interaction with the environment). Millikan's theory, therefore, is both highly complex (Buller 1999a) and ingenious (Griffiths 1993).

6.4 The Continuing Usefulness Requirement

A trait type *X* has the proper function *F* (at time *t*) if

(i) *X* has arisen, been modified, or been maintained by natural selection at some point (before *t*) because its doing *F* contributed to fitness of individuals with *X*; and

(ii) *X*'s doing *F* has recently and importantly (before *t*) causally contributed to the survival and reproduction of organisms in this species with this trait.

What this account accomplishes is that it explains how traits can lose their proper function. The eyes of cave-dwelling species (Griffiths 1992, 1993) can lose their proper function of seeing if they no longer use their eyes to see or if their seeing no longer contributes to survival and reproduction. The "recently" refers to numbers of generations rather than years to explain differences between species. The vagueness of "recently" here is similar to Godfrey-Smith's (1994) "the recent past" and Griffiths's (1992, 1993) "evolutionary significant period." The word "importantly" is to exclude a single event or something equally trivial restoring a vestigial trait.

Schwartz emphasizes that the notion of cause in (ii) is to be construed as permissive; it is meant to allow a wide range of ways that a trait may contribute to survival and reproduction. Take the flight response associated with the emotion of fear as an example. This response, biologists suppose, was selected for because it enables organisms to escape predation (and so survive and reproduce). But it could also be argued that the flight response also assists in tuning the organism's musculature through fright-induced exercise and minimizing intravenous atheroma buildup, again through fright-induced exercise. Many effects can count as putative contributions to survival and reproduction but (i) eliminates most from counting as proper functions. Thus, if biologists believe that the function of the emotional flight response is to tone musculature and thus contribute to survival and reproduction, it cannot be a proper function unless biologists also believe that natural selection favored this effect. They do not. They have no doubt that natural selection favored the flight response to enable the avoidance of predation (and other dangers) because the fear response appears to be designed to avoid predation (Dawkins 1986).

The continuing usefulness requirement avoids the "substantial biological commitment" of a strong adaptationist construal of proper function. It includes a weak adaptationist perspective (Schwartz 2002, 256) by insisting only that "*some* aspects of *some* traits" are proper functions.

6.5 The Biosemantic Theory of Mental Content

Biosemantic theories of mental content are naturalistic, teleological theories of mental content. They are naturalistic in that they attempt to explain what it is about mental representations that enable them to represent something in terms consistent with those of the natural sciences. These theories typically include picture theories, causal or covariation theories, information theories, and functionalist or causal role theories. They are teleological (from the Greek *telos*, purpose, function, goal), as previously noted, in that they refer to the biological function or purpose of intentional mental states. The teleological component of the theory suggests that true representations are produced when the system housing the mental representation is functioning properly, that is, functioning in accord with either natural selectionist history or selection through learning (Millikan 2009).

The brain was and is designed by evolution or learning, in interaction with a range of heterogeneous resources including other brains, to make or learn to make representations.[2] Teleological theories thus sharply differentiate intentional signs from natural signs. Even when intentional representations are true, neither the fact that they represent nor what they represent is entirely determined by any current relation they actually bear to their representeds. The representational status and the content of the intentional representation are both partly determined by reference to its natural purpose or the natural purpose of the biological systems that produced it, and these purposes are determined by history, by what the mechanisms were selected for doing, either during evolution of the species or through earlier trial-and-error learning (Millikan 1984). Naturalistic teleological theories of content, therefore, are externalist insofar as they take beyond-the-individual factors (i.e., evolutionary history) as importantly relevant in the metaphysical determination of mental states. In this they differ from individualist (or internalist) theories, as discussed previously. I return to this subsequently, to discuss the "mark of the cognitive" intracranialist objection.

I have one more final point to make concerning teleological theories: they are based on the notion of function in selection, and this implies that they must explain how the production of representations can sometimes benefit an organism. This in turn implies that the organism must use them in some way; and this implies that there must be part of, or some activity of, the organism that interprets or understands them (Millikan 1984).

According to biosemantics, several different kinds of processes use representations, from the relatively simple (e.g., animal signaling systems) to the highly complex (e.g., human doxastic systems), two of the most important being descriptive-indicative and directive-imperative. The former are designed to represent facts, and the latter to tell what to do. The most primitive type of representation is designed to do both, to describe what is the case and direct what should be done about it. Millikan (1996) terms these pushmi-pullyu representations and cites the dancing of honeybees as exemplary.[3] The dance of the honeybee tells where the nectar is and simultaneously where the watching bees must go to find it.

Representations are often considered rather more sophisticated than the primitive pushmi-pullyu suggests; for this reason, Millikan (1984) prefers the term "intentional icon" to refer to all representations, from the most primitive pushmi-pullyu to those of considerable sophistication that are included in human inference processes.

Each type of intentional icon includes a producer and an interpreter component. Some producers and interpreters will be within the one organism (e.g., the producers and interpreters of hormones), and others will be in separate organisms. The honeybee dance, for example, is produced in the dancing bee and interpreted in the watching hive bees. The important point is that what the producer does helps the interpreter to perform functions that are mutually beneficial in either increasing the likelihood that both will survive and proliferate or maintain their current settings (selection through learning). The presence of both producer and interpreter is part of a normal mechanism by which the other helps itself to survive or proliferate, and this cooperative interaction is the result of past selection or learning that operated on both simultaneously.

Producers produce intentional icons that will be true (indicative-descriptive) or satisfied (directive-imperative) only if they map onto some state of affairs in the world in accordance with some sort of "semantic" rules. "These are rules of correspondence between signs and world affairs that have been instantiated in the past when the consumer and the producer or their ancestors have succeeded in performing their cooperative functions" (Millikan 2009, 397–398).

The semantic rule associated with a descriptive intentional icon determines the state of affairs that must obtain if the consumer is to perform its tasks, whatever they may be, in the normal way. The consumer varies its activities systematically according to the icons presented to it to ensure that its activities conform to, or are *adapted* to, the state of affairs represented by the icon so that it performs its function properly given the state of affairs. Indicative-descriptive icons are designed to stand in for states of affairs typically outside the organism involved and to vary systematically with such states, governing behavior, both internal and external, to accord with or adapt to such states of affairs.

It is clear that from this account that the bee dance is, in part, a descriptive intentional icon. If the bees are to perform their function by locating nectar by responding to the dance in the normal way, the dance needs to correspond by a certain semantic mapping rule to a fact about the location of nectar. An indicative intentional icon will be true (represent truly) only if it represents according to a certain semantic mapping rule.

Directive-imperative intentional icons that correspond to mapping rules are "satisfied" rather than "true." The semantic rules associated with directive-imperative intentional icons determine the state of affairs its interpreters are to produce. It is the function of the interpreter to bring it about that a state of affairs corresponds by rule or *adapts* to the icon. The bee dance, therefore, is also in part a directive intentional icon because it must correspond by a certain rule to the direction of flight of the hive bees if their dance-interpreting mechanisms are to serve their functions normally.

Drawing heavily on Millikan (2009, 398) here, I provide emotional ontogenesis as an example. The activities of infant and caregiver fulfill their cooperative function of directing the caregiver to provide the requisite assistance in the normal way, that is, through the characteristic mechanisms that have accounted for previous success and subsequent selection of the assistance-soliciting and assistance-providing devices. The infant expresses a need that maps, by a certain rule, first onto the nature, location, intensity, and so on, of the need, and second onto the nature and duration of the assistance provided by the caregiver. In each case, had certain variables of the assistance-soliciting expression been different (e.g., tone of cry, facial configuration, motor activity, etc.), the need or deficit would have needed to be different in type, location, and so on, and the caregiver would have needed to provide an alternative type of assistance for the expression to serve its purpose in the normal way. The semantic rules are defined by the way in which the set of possible needs, deficits, and pleasures determines an isomorphism between these producer-interpreter domains, an isomorphism that holds between the set of possible emotional expressions and the set of responses that will result if caregivers provide assistance through historically normal mechanisms.

Mappings or isomorphisms are an essential feature of each kind of sign system. Intentional icons are produced by systems that are designed to produce them to accord with predetermined mapping rules to which their interpreters are adapted.

It is worth noting that there is no limit to the complexity of the semantic mapping functions that can map intentional icons onto states of affairs they represent. "A bizarrely coded secret message from a CIA agent can be as much an icon that maps on to a certain world affair in accordance with a definite semantic-mapping function as any bee dance, sentence or diagram. Intentional icons must be things apt for use by icon users, but icon users can be very idiosyncratic in their habits" (Millikan 2009, 309). Every representation is in some kind of code, but the complexity of the code is irrelevant.

In Millikan's account, biological devices are the most familiar items with proper functions. The various traits of biological items that are members of reproductively established families have been selected for by Darwinian natural selection. She emphasizes, however, that artifacts and cultural practices and traditions (including linguistic traditions) all have functions in her sense of proper function. In addition, the performance of a device's proper function almost invariably depends on it having a suitable surrounding environment that contains members of other reproductively established families that need it, in turn, to perform properly. There are three categories of proper function germane to the present analysis; these are relational, derived, and adapted proper functions, and it is to these that the discussion now turns.

6.5.1 Relational Proper Functions

Having relational proper functions is typical of the behaviors of organisms. These functions in higher species, controlled mostly by perceptual and cognitive systems, are performed by altering the relation between the organism and the environment as required so that the environment will supply surroundings and inputs conducive to the organism's needs. Some functions involve changing the environment to suit the organism, some involve changing the organism to suit the environment, some involve merely changing the relationship of organism and environment, and some involve some combination of all three (Millikan 2002, 126). The devices responsible for these changes have relational proper functions of one kind or another; that is, their function is to ensure that environment and organism bear some particular relation to each other. Their function is to create relational structures. As with any other proper function, a relational proper function of a device corresponds to an effect that ancestors of the device have historically possessed that assisted in accounting for its selection. In this case, the function was to create an abstract relational structure.

Millikan's favorite examples are the chameleon and the honeybee (1984). The chameleon is equipped with chromatophores that contain granules of pigment that, through aggregation or dispersion, change the color of the chameleon's skin to match that of its background (Cott 1966). The chromatophores' function is to protect the chameleon from predation, and they accomplish this by producing a skin color that bears the relation *same color as* (Millikan 1984, 35) to the chameleon's background environment. To create this relational structure, the device causes changes in the chameleon but not in the environment. To produce the relational structure *skin bearing same color as its background,* the chameleon does not need to produce both relata. It produces only one relatum, the skin color, but in such a way that it creates the requisite relation.

As for honeybees, their dance-producing devices have the function of producing a relational structure with two relata, the location of the nectar being one relatum and the particular dance produced another. The relation in question is the one given by the function (in the abstract mathematical sense) describing the semantic rules of "beemese" used by this particular species of bee. In most species, an angle that the dance movement marks out relative to aspects of the hive always bears the same definite relation of the angle of the location of nectar relative to hive and sun. The properly functioning dance-producing device always produces exactly the same thing, namely, this requisite relational structure. The device that produces the dance has an additional proper function, that is, to send hive bees flying to the nectar source. The relational structure thus produced has as relata (a) the orientation of the dance relative to the hive, and (b) the angle in which hive bees fly relative to hive and sun. The producer devices create this relational structure by altering the environment; there are no changes required in the dancing bee itself, but only in the hive bees. As a result of producing the proper dance–nectar location structure, the properly functioning device later produces always exactly the same thing, namely, existence of this second relational structure between angle of dance and angle of flying.

In short, where a proper function of a trait is to produce a series of effects, each affecting the next, it is also a proper function of each stage to produce the next. In honeybees, this results in the production of one relational structure, *dance–nectar location,* having the proper function of the production of another, *dance–direction of flying.* Finally, and as a result, yet a further relational structure is to produce *hive bees flying toward nectar.* This is typical of biological items. Proper biofunctional relational structures typically fulfill their functions by producing further relational structures that eventually produce a relation between organism and environment, yielding conditions or inputs conducive to the organism's well-being.

In sum thus far: It is typical in biological items that proper functions involve series of interweaving stages, each of which is an abstract relational structure, some moments in these processes producing changes in the organism, others producing changes in the world, but involving always exactly the same relations, although among different relata each time they are run. It was to simplify these complex relational structures that Millikan (1984) introduced the terminology "adapted" and "derived" relational proper functions (2002, 128).

6.5.2 Adapted Proper Functions

Adapted proper functions are relational proper functions adapted to particular contexts. The chameleon and honeybee are again instructive. Imagine that the chameleon sits on a green background; the adapted proper function of the chromatophores in this particular case is to turn the chameleon's skin color green. This function is adapted to a given context (Millikan 1984, 40). Similarly, when the searching bee locates nectar three kilometers north-northeast of the hive, her dance will map this specific location in accordance with the "semantic rules" of beemese (and acquire a further adapted function by sending the hive bees off in a specific direction and to a specific distance, that is, three kilometers north-northeast of the hive).

The configuration *chameleon bearing the relation same color as* what it sits on is a member of a higher-order reproductively established family (because chromatophores in the chameleon proliferated because of their function to do so), and so too is the bee dancing *thus to the location of some nectar* (Millikan 1984, 40), and for the same reason; but neither the green skin of the chameleon nor the bee's dance of three kilometers north-northeast is (because the devices that produce these specific color and dance features did not proliferate because they produced these specific colors or dances). The green color and the specific bee dance are adapted to some other item, that is, the specific background color green and the specific nectar location; Millikan terms these devices *adaptors* (1984, 41). A derived proper function is a function derived originally from the relational function of the producing device plus its context, in these examples from the adapted functions of the chromatophores and the dance-producing device as operating in a certain context. All derived proper functions are adapted in this way.

Consider any device that has as a function the production of any kind of learning on the basis of experience, for example, a device that effects trial-and-error learning, learning by imprinting or imitation, or even learning by reasoning to a conclusion from premises derived from experience. Any such device has a relational function or functions. It is designed to turn out behaviors (e.g., beliefs or desires) as a function (in the mathematical sense) of certain designated kinds of input from experience. Baby sea otters learn to eat whatever their mother eats; their producer devices therefore have a relational function to create in the baby otter precisely this disposition. Given that the mother otter eats sea urchins, then, it is an adapted function of the device to produce a state that effects a disposition in the baby to eat sea urchins. And once this complex state is in place in the baby otter, it has a derived proper function to produce sea urchin collecting and eating.

It is the case that many biological devices have proper functions that they perform by varying (i.e., adapting) their performance in accordance with present conditions (their adapted device adaptors) to produce an invariant end result, such as not being eaten or securing a supply of nectar. In contrast to the chameleon and honeybee, the human neonate is not completely genetically equipped with adaptor devices, that is, devices whose relational proper functions are to adapt the assistance-soliciting and assistance-producing devices to particular features of a given context (e.g., green background, nectar at three kilometers north-northeast). The human infant must construct nearly all such devices.

In terms of human emotional ontogenesis, there are a number of important points to note at this juncture:

(i) The assistance-soliciting mechanisms in the infant and the assistance-providing mechanisms in the caregiver(s) are coconstructed or learned through linguistically mediated social interaction. The inbuilt interoceptor and proprioceptor devices in the neonate trigger the referentially opaque precursor expressions in infants, such as unfocused crying, motor unrest, and so on, and these trigger the inbuilt nonspecific general-purpose interpreter devices of intuitive parenting skills in caregivers (to search, identify, and correct needs or deficits); these repeated search processes serve to fine-tune the interpreter devices to the infant's idiosyncratic needs, causes, and satisfaction conditions. These are fed back to the infant in linguistically mediated social interaction to promote the maturation of the neural substrata of specific producer devices, such as sadness in the infant. What this implies is that, in contrast to inbuilt producer-interpreter devices in honeybees, the emotionese producer-interpreter devices in human beings are largely learned through social interaction. Emotionese will, of course, include a limited range of inbuilt emotion elicitors (e.g., fright response to precipice stimuli), but it will also include a whole raft of culturally relevant and culturally appropriate elicitors (e.g., blue exam books and summonses from senior colleagues) and responses and culturally relevant linguistic labels for these elicitors. In Indonesia, for instance, crawling infants elicit disgust because they look like four-legged animals, which Indonesian culture despises. Other cultures have variants of what in English we term joy, such as *amae* in Japan, a feeling of pleasant dependence on individuals or institutions (Morsbach and Tyler 1986), and what in English we term anger, such as *song* in Ifaluk, which is an anger analogue but is more akin to simple admonition with no disposition to revenge (Lutz 1986). These emotions or emotion terms refer to responses that are entirely appropriate to the host culture in which they are embedded. Japanese lifestyles are such that dependence is viewed positively, and given the hostility of their natural environment, aggressive, retributive anger would be massively counterproductive for groups of Ifaluk.

(ii) Human emotions possess the relational proper function of producing reactions that bear a specific relation to environmental saliences concerned with well-being (and thereby ensuring social regulation and harmony). The specific emotions are adapted to particular emotion-eliciting stimuli that are frightening (e.g., spiders), saddening (e.g., loss of Teddy), pleasing (cuddles), and so on. Such stimuli are the adaptors of specific emotions. Each specific emotional producer device has an adapted proper function to produce, as a function (mathematical sense) of the semantic rules of human emotionese, the relational structure *response tailored to specific environmental stimulus*. In addition, these adapted proper functions have further derived proper functions to produce, again as a function (mathematical sense) of the semantic rules of human emotionese, relational structures of even more specificity, for instance, to produce the relational structure *bear approaching–move away*. These adapted and derived proper functions act as adaptors for further adapted and derived proper functions as complex capacities are functionally decomposed in detailed functional analyses. They are reflected right through functional networks, terminating, as they must, in the brute neurochemical devices that eventually produce the most basic effectors, such as neuronal depolarization and activation of IEGs.

(iii) Yet another point to note is that the redirection of the imperative icon from other to self indicates the development within the emoting subject herself of her own internal interpreter devices. These too are gradually acquired in interaction with caregivers, simultaneously with producer devices. Both the interpreter and producer devices of the referentially opaque pushmi-pullyu "Something's wrong! Do something!" are gradually elaborated and refined in light of, and simultaneously with, feedback from the caregiver's progressively refining search-and-correct activities. "Something's wrong! Do something!" is gradually refined to more specific pushmi-pullyus, "I'm cold, cover me!" "I'm frightened, comfort me!" and so on. This too is evident in the vocalizations of young children. Around two years, at about the time of the word spurt, they begin increasingly to use and refine their own emerging interpreter devices by mirroring caregivers' earlier utterances (Holodynski and Friedlmeier 2006). Instead of the caregiver saying, "Amy sad," "Amy lost toy," and so on, Amy herself begins to comment on her own feelings (provided, that is, that they are not too intense; see chap. 4), their apparent causes, and what to do about them. Indeed, young children learn language initially by learning emotion words, for example, "Amy sad," "Tommy love Mummy," "Teddy happy." This is consistent with what has been consensually acknowledged for many years, that is, that human neonates are also preadapted to acquire language in sociolinguistic interaction. In an important sense, therefore, human children learn to emote as human beings as they learn to speak emotion words. As one theorist (Shanker 2008, 70) observes, "emotions appear to serve as the very architect of language development."

In addition, the emergence of some voluntary control indicates a gradual if incomplete separation of indicative-imperative functions. The indicative-imperative cleavage becomes more apparent as the higher cognitive emotions begin to develop from about four years of age. The cleavage is complete in adult emoters, who can reflect calmly on the causes and consequences of at least some of their emotions, "relive" them (from memory), and decide how to deal with them in the future. ("I am embarrassed; I talked too long and too loud today; next time I'll talk less and more quietly.")

All of this is possible because emotion precursors and caregiver intuitive parenting skills are devices that possess the proper function to learn on the basis of experience; they are to produce behaviors, such as imitation, trial-and-error experimentation, even beliefs and desires, as a function (mathematical sense) of certain kinds of experiential input. The auditory, visual, tactile, and olfactory sensory-perceptual stimulation provided in neonate-caregiver mutual feedback provides (most of) this experiential input. The inbuilt emotion precursors and sensorimotor preadaptations in the neonate, together with the caregiver's inbuilt intuitive parenting skills, form the basis for the development of full human emotionality given these experiential inputs (i.e., the normal conditions for proper function). The interaction-coaction cycles in which both parties engage, which progressively lead to the development of the exquisitely fine-tuned emotional devices in both parties, are the result of mutual adaptation of relational proper function.

The devices in both neonate and caregiver possess proper functions to produce, first, dispositions to imitate; then, second, imitation dispositions have the relational proper function to produce, in both parties, the disposition to do whatever the caregiver (or neonate) does, which has the further adapted proper function of producing the disposition to imitate the smile, frown, noise, and so on, that the other is producing, and then the further derived proper function to produce this particular imitation behavior. The neonate's producer devices are adapted to the caregiver's adaptors, and the caregiver's producer devices are adapted to the neonate's adaptors.

Both neonate and caregiver learn through feedback which emotional stimulus is causing the infant to (precursor) emote and what to do to remedy it. The inbuilt caregiver producer devices have the relational proper function to create in the caregiver the disposition to search-identify-correct, and once this disposition is created, it has the derived proper function of producing the requisite search-identify-correct behaviors. In due course, through trial-and-error experimentation (informed by general cultural ethnotheories, emotion-specific ethnotheories, and her increasing knowledge of her infant), the caregiver learns precisely what causes her infant's distress and how to remedy it. This is because her producer devices have the relational proper function to produce dispositions and then behaviors that correlate with positive feedback, such as smiling, cooing, and cuddling up. This process also operates in the child, who progressively complements imitation with trial-and-error learning (as these producer devices are gradually constructed), increasingly scaffolded by language (from about twelve months), and eventually reasoning from premises derived from experiential input to conclusions (again, as these producer devices are gradually created).

In addition, the neonate's responses of smiling, cooing, and snuggling up, as noted, are accompanied by feelings of great pleasure; the responses are imitated by the caregiver, and the feelings of great pleasure that attend them are experienced. These mutual responses and feelings form the basis of the special social relationship that develops between caregiver and infant.

As the caregiver goes through her baby-talk-commentary-accompanied search-identify-correct routine, she provides the necessary visual, auditory, tactile, and olfactory stimulation to excite the activity of the infant's intracellular IEGs (and possibly a host of other genes). The relevant IEGs respond by developing, maturing, and connecting neural circuits that will eventually underpin the child's discrete, fine-grained producer and interpreter devices. This too reflects an adapted relational proper function where the release of specific IEGs is adapted to particular sensory-perceptual stimulation. Again, in exactly the same way, the interaction with her infant provides the necessary sensory-perceptual stimulation to excite the maturation of the caregiver's neural circuits that eventually underpin her fine-grained interpreter and producer devices. In other words, the operating IEGs, which are members of first-order reproductively established families, gradually construct emotional neural substrata that are members of higher-order reproductively established families.

(iv) As basic emotions and higher cognitive emotions develop, they are experienced consciously by the individual emoter-subject (child or adult). A range of subpersonal biological processes that extend across the body-environment boundary underpin this conscious experience, but the emotions are felt at the personal level. A range of subpersonal transcranial processes are implicated in emotional ontogenesis, experience, and expression, but no set of neurons or neuropeptides or any other subpersonal device emotes. Only individual animals emote. This seems entirely appropriate; the emotions evolved to protect the interests of whole organisms, typically by movement away from noxious stimuli in flight, or in the cessation of movement in freezing. An additional point is that whole organisms are typically required to provide the necessary background conditions for the efficient functioning of component systems (see chap. 3).

(v) The foregoing process implies no infinite regress. The producer devices in the infant are activated initially by homeostatic deficits, and even if interpreter devices are required to read them off, these deficits are grounded in physiology. Similarly, at the opposite end of the process, further interpretation is not required, as the process terminates in actual behavior in the world. In sum, human emotions have as a direct proper function to perceive and respond appropriately to environmental saliences connected to well-being. This is their invariant end result. This invariant end result, however, is achieved through a series of adapted processes, the adapted proper functions of which are to respond appropriately to specific stimuli that are frightening, saddening, frustrating, interesting, pleasing, and disgusting, and so on, which are the adaptors to which these are adapted.

6.6 Natural Signs and Intentional Signs

I have noted repeatedly in previous chapters that positive and negative stimulus thresholds are detected in human neonates by a range of surface sensory receptors (which detect bright lights, loud noises, physical discomfort, etc.) and internal interoceptors and proprioceptors (which detect saline and glucose depletion, changes in facial expression, etc.). These receptors activate, via the appropriate thalamic projections, the relevant nuclei in the amygdala, which identify stimulus positivity and negativity in terms of well-being. The amygdala circuits, in turn, activate the appropriate hypothalamic circuits, resulting initially in motor unrest and unfocused crying (and progressively in the full suite of emotional reactions once the requisite neural substrata are developed). Unfocused crying and motor unrest are recurrent natural signs of homeostatic distress; that is, they carry local natural information concerning their bearer's homeostatic well-being to (a) consumer device(s). That they do so, however, is no accident; the amygdala and associated structures evolved precisely to do this. There is, in short, a real causal connection between the state of affairs signified (saline or glucose depletion) and the resultant sign (motor unrest, crying). Insofar as the amygdala and associated structures are devices selected for this function, the unfocused crying and motor unrest elicited are intentional signs (icons) of homeostatic deficit.

Ethologists suggest that communicative animal signals appear to have evolved from preparatory movements, called "intention movements," made at the onset of some activity (Millikan 2004, 103). These movements originally served as cues to conspecifics of either the animal's ensuing action or the stimulus that excited such activity, but only accidentally. They served as natural signs for danger, food, readiness to mate, and so on. When it became advantageous for conspecifics to know what the animal was preparing to do, and what it was doing it for, the intention movements gradually became genetically enabled, stereotyped, and exaggerated. Similarly and simultaneously, conspecifics interpreting the sign gradually became tuned to respond to the sign appropriately by learning and genetic selection. That conspecifics are aware of what one of their number is preparing to do, and why, is often useful to animals. For instance, most basically, it is useful for an animal to know when its partner is ready to mate.

Intentional signs thus arose from natural signs quite rapidly through a ratcheting process (Tomasello 1999), facilitated and expedited by learning, through imitation and concept formation on the part of human beings. It is only a small step from a nonlinguistic intentional sign to a linguistic intentional sign if one is genetically so predisposed. Natural signs, intentional signs, and linguistic signs all carry locally recurrent natural information (the kind of information that inhabits tight worlds).

An example: snow geese flying north over Whyte Avenue in Edmonton, Alberta, in April is a natural sign that spring will shortly arrive. The arrival of the geese carries natural information local to Alberta, Canada. Snow geese flying north in April means that spring will soon follow because, in a statistically significant number of cases, the arrival of the geese has been followed by the arrival of spring. It is because there is a causal connection, that is, a nonaccidental statistical correlation between states of affairs signified and sign or signal, that local natural signs are recurrent, that is, they carry locally recurrent natural information. Real causal connection is the reason that many states of affairs persist and spread and replicate themselves over space and time and are experienced as occurring regularly, repeatedly, and often cyclically. The niches that animals inhabit are "tight"; the same objects, and the objects to which these objects are related, are experienced from a range of perspectives with mundane regularity, that is, as locally recurring. This locally recurrent information and the natural signs that carry it are what enable animals to learn about Bs (states of affairs signified) from encounters with As (signs or signals). Correlations are defined relative to reference classes (Millikan 2004, 37), and the relevant reference class will include all those "samples" to which the correlation of As and Bs in the present sample are projectable.

Views differ on what counts as natural information. Dretske (1981) insists that a sign or signal carries information about a state of affairs (source) only if the signal's occurrence yields a probability of 1, determined in accordance with natural and logical necessity. Other theorists, including Millikan (1984), Price (2001), and myself, disagree (and even Dretske [1986, 1991] seems to equivocate); we suggest that a sign or signal carries information as a matter of statistical frequency, not natural law. This is partly because in nature there is no natural law that precludes the evolution of new species or devices that cause the same response. No natural law, for instance, precludes new eliciting stimuli from causing the fear response in emoters in any given culture. However, it is mainly because virtuous cognition involves the trade-off of two types of correlation, not just one, as Dretske believes (Godfrey-Smith 1991).

By using signal detection theory and Field's (1990) construals of head-to-world and world-to-head reliability correlations, Godfrey-Smith (1991) persuasively argues that virtuous cognition involves a trade-off between the two, depending on the relative significance of signal to noise (stimuli that falsely activate inner-action-guiding states).

A perfect head-to-world correlation exists when an inner state is tokened only if certain external conditions obtain; it is not required that the inner state always be produced by the external condition. A perfect world-to-head correlation exists when an inner-action-guiding state is always tokened by the external condition, whether or not the inner state obtains under other conditions.

"Except in the most epistemically benign and friendly environments, the two kinds of reliability cannot be bought as a package deal; rather, one must be traded off against the other" (Godfrey-Smith 1991, 718). The point is illustrated in extreme cases; one way to achieve excellent head-to-world reliability is to token action-guiding states, but this sacrifices nearly all an animal's world-to-head reliability. Excellent world-to-head reliability can be achieved by activating the action-guiding inner state in response to all forms of noise, as well as correct stimulation, but this leads to too many false positives. These two types of reliability correlation must be traded off in virtuous cognition depending on the particular situation the animal finds itself in. The reliability trade-off depends on the degree of probability (which depends on stimulus or noise significance) and the costs and benefits differentially associated with "hits" (correct positives), misses (false negatives), false positives, and correct negatives. The more actual survival depends on hits, that is, as the importance of signal relative to noise increases, the higher will be the trade-off of head-to-world reliability in favor of world-to-head reliability.

An example: A rabbit and a fox find themselves in the same field, and both notice movements in the nearby grass. The fox either interprets or fails to interpret the stimulus as prey; if he correctly interprets it as prey (scores a hit), he expends energy to capture it and benefit. If, however, he "misses" the prey, it costs him nothing in terms of energy expenditure. Compare the rabbit: if she scores a hit by interpreting the stimulus as predator, she expends energy to escape. If she misses, however, she could be killed or injured. The stakes for the rabbit are so high that she sacrifices head-to-world reliability to obtain an extremely high value of world-to-head reliability and tolerates a number of false positives to achieve as many hits as she can.

Other points relating to natural signs are worth noting. The first is that natural signs are structured (as, indeed, are the states of affairs they signify). It is not geese flying south per se that is a natural sign that spring will soon arrive; rather, it is geese at time T flying over a certain place P (Whyte Avenue) that is a sign that spring will arrive soon after T at P (Millikan 2004). It is the relational structure of a natural sign that determines its meaning. The meaning of a sign is determined as a function of values of significant variables as determinables manifested by the sign. The meaning varies systematically to parallel significant transformations of the sign. What this implies is that a natural semantic mapping is recurrently applied.

The second is that the meaning of public language is determined by compositionality, but in the account given earlier, compositionality is just one way among many of being relationally structured (Millikan 2005). Representations that exhibit compositionality are transformed into other representations in the same system by rearranging the same parts or substituting parts for others. It is not compositionality in particular but relational structuring more generally that yields the productivity of sign systems, their capacity to yield novel information or ways to convey information.

What this implies is that what recurs in the case of recurrent signs are other members of the same system of signs, where a sign system is composed of a set of possible sign types, which are transformed (in an abstract mathematical sense) systematically by rule to correspond to transformations in what they signify. Time and place variables in any structure can be transformed by projecting forward in time and east or north in place, for instance, by observing another type of migrating goose flying north over the Yukon in May; by using precisely the same transformations, another (possible) natural sign in the same sign system can be produced. In short, recurring natural signs are semantic mapping functions from signs to signifieds that define isomorphisms between the set of possible signs in any sign domain and their set of possible signifieds.

The third is that different natural signs can signify the same, more distal state of affairs by a different route or routes that are only partially the same. For example, certain patterns impinging on my retina at a certain time and place are local signs of the occurrence of the shape, color, size, and so on, of my crying infant that signifies a prior absolute stimulus threshold deficit, and these patterns are routes to the more distal state of affairs of, for example, glucose or saline depletion. The sound of certain sound patterns emanating from other family members ("it is four hours since her last feed") and impinging on my auditory apparatus signifies what the crying means (absolute stimulus threshold deficit) and is also part of the route to the distal state of glucose or saline depletion. What this implies is that natural signs typically co-occur with other signs in the relevant domain of reference.

As already indicated, local natural signs of more distal states of affairs carry local information about all the more proximal states of affairs en route from them to those distal affairs (e.g., an infant is crying because she is hungry, wet, startled, etc.). However, a natural sign that is also an intentional sign will carry only some of this information intentionally, and this is the information it was selected to carry to be useful to its interpreters-consumers. The information may be very abstract and may relate to very distal affairs, but this matters not; the only information that matters is that which is of use to interpreters-consumers. What this means is that if interpreters-consumers were designed to use only that *food is near* or that *Amy is hungry*, then it need not carry any of the more proximal relevant information such as patterns on the retina or sound patterns in the auditory apparatus. Percepts are designed to carry information about a distal causal antecedent without apparently carrying information concerning the more proximal members of the causal chain (Dretske 1981). The percept treats proximate members as transparent to represent its more distal causal antecedents. The only information a percept carries intentionally is the information it was selected for carrying, that is, only the information that is used by its cooperative interpreter.

Consider an individual, Mary, who sees an intruder, armed with a knife, enter the bedroom in which she has been sleeping. Certain patterns on Mary's retina that signify the presence of the intruder mediate between Mary and Mary's awareness of him. She will not possess an intentional representation, however, of the retinal patterns that recognize the intruder. In addition, certain patterns traversing the optic nerve will typically signify what is happening on the retina, but they will not be intentional signs of this. An intentional sign signifies only what it is used to represent when it is functioning normally. The patterns on Mary's retina are not used to guide Mary's behavior toward the retina or any inferences about the retina. It is irrelevant to the proper functioning of these patterns that they have been produced in the normal way from retinal images as long as they are consistent with the intruder. The images on the retina are of absolutely no concern to Mary.

Exactly the same is true for public language (linguistic) intentional signs. The English sentence "The sun is shining" is a recurrent natural sign that the speaker believes that the sun is shining. Its memetic function, derived compositionally from the combined memetic functions of its signified components, however, is to produce beliefs in hearers that the sun is shining, not that the speaker believes that the sun is shining.

Like locally recurrent natural signs, intentional signs are always members of a domain exemplifying a system of probable signs, the entire system running isomorphically to the domain of signified states of affairs. There is no need to restrict the complexity of the semantic mapping functions that might map representations onto their representeds. Signs must just be useful to their consumers, and these can vary enormously in nature. Every representation is in some kind of code, but the complexity of the code is irrelevant. The only factor that is relevant is the ability of the consumer to use them.

6.7 Linguistic Signs

Linguistic signs do not represent states of affairs directly; they intentionally represent properties that are natural signs of these states of affairs. Linguistic signs accomplish this, however, only when they are functioning properly, that is, functioning in the way that accounted for their selection and proliferation in a language community. To function properly, the properties they represent must actually be natural signs of something. Linguistic signs can also represent other representations as representations, and the repeated representations may or may not be signs of anything real.

6.8 Meaning and Its Acquisition

Meaning is generally construed to be "the conventional, common or standard sense of an expression, construction or sentence in a given language, or of a non-linguistic sign or symbol" (Loar 1995, 471). Emotions and linguistic forms are all intentional icons that differ simply in terms of relative complexity and sophistication and proximality or distality, and as I have just argued, human emotions (and their precursors) are intentional signs carrying locally recurrent natural information. Prelinguistic emotional meaning is a function of genetic endowment, and linguistic meaning is a function of linguistic convention and conception at three different levels that dictate how sentences and utterances are standardly constructed and interpreted.

As I have already indicated, the overall purpose of human emotionality is social regulation and control, whereas the purpose of natural, public language is to enable communication between persons in the same speech community. Human language evolved, as did nonhuman calls, grunts, and gestures, to enable communication between conspecifics. As the size and foraging range of human social groups increased, full human language evolved from protolinguistic precursors (Wray 2000, 2002), initially to track present and past social activities and predict future ones. The evolution of full human language complemented and extended the role of prelinguistic emotion in early humans to keep track of who is (was, will) be doing what to whom, where, how, for what reason, and with what results (Dunbar 1996). Subsequently it was exploited to allow the verbal transmission of object and event description and complete causal mythologies about the world. Even more recently, language also enabled the sharing of incomplete ideas and causal theories leading to the refinement, elaboration, and revision of ideas, that is, knowledge construction (D. Johnson 2003). Both these functions of language are required to enable human beings to survive and flourish in increasingly complex physical and social environments.

For any communication, prelinguistic or linguistic, to be possible, however, (i) the interactions of communicators or interlocutors must be coordinated; (ii) the contents of their expressive behaviors, including their utterances, must map words relatively faithfully to objects and events in the world and otherwise comply with "satisfaction conditions" (discussed below); and (iii) communicators or interlocutors must possess the appropriate perceptual and conceptual structures to enable the referential grasp of objects and events in the world.

(i) Each sign system, including natural language, is characterized by a public pool of conventional sign forms, each with its own conventional function, that is, facial expression, vocalizations, body orientation in terms of emotion and words, syntactic forms, tonal inflections, idioms, and so on, in terms of public language. These are what Dawkins (1976) terms "memes," that is, cultural artifacts or items that are reproduced by imitation rather than genetically. Popular tunes, dress fashions, ideas, values, forms of emotional and linguistic expression, and so on, are memes, and they are copied from one person to another and from one generation to another. Dawkins reminds us, however, that human memory and energy are not inexhaustible; this means that memes have to compete with each other for reproduction. Memes that are selected for reproduction are selected for their purpose; emotional and linguistic memes are selected for their cooperative purpose, and

> the mechanisms that produce them have evolved to promote social coordination (Millikan 1984). Coordinating functions are served when the purposes of two or more individuals are simultaneously facilitated, bringing the separate behaviors of those individuals into a designed coordination that benefits both but has not been designed by either individually.

Emotional expressions, such as facial expressions and display rules, are reproduced from one generation of emoters to another and reproduced in the same subject-emoter on different occasions. The emotional responses and expressions are reproduced in specific emotion-related contexts simply because of the effects that they have on interactors in some critical proportion of cases. These effects are precisely those the emotions evolved to produce and subject-emoters wish to produce. In a similar fashion, natural-language devices, such as words and phonemes and occasionally sentences, are reproduced from one generation of speakers to another and reproduced by the same speaker on different occasions. It is clear, however, that speakers reproduce words in specific concrete syntactic patterns rather than in others simply because of the effects these patterns have on hearers, in some critical proportion of cases. These effects are precisely the ones that speakers wish to produce.

The proper effects on interactors in an emotional context to those emotional expressions—the proper reactions, if you will—are also reproduced. This is because interactors benefit (in terms of information received, rewarding feedback elicited) from reacting properly to the expressions they witness. Specific emotional reactions produce interpretations and responses in interactors in conformity with the appropriate mapping rules, and these are often enough to ensure that reproducing a reaction provides interactors with rewards in terms of information or action.

The proper effects on hearers of these language forms (the proper reactions, if you will) are also reproduced. This is not due to hearers reproducing each other's reactions, however, at least not typically. It is due to the fact that hearers benefit from reacting properly to the sentences they hear. Words in certain concrete syntactic patterns produce beliefs in hearers and actions in conformity with the rules of language, and these, often enough to ensure reproduction of reaction, provide hearers with rewards in terms of useful information or action.

Memetic linguistic forms are real (although what they refer to may not be real, e.g., the tooth fairy). They exist in the coordinated utterances of speakers and hearers (and their "private" utterances and thoughts). Words and meanings are produced by precedent, and they coordinate the interactions of interlocutors. The conventional function of such linguistic forms will survive for only as long as they serve the interests of both speakers and hearers often enough to avoid extinction. The proper function of any linguistic device that accounts for its continued survival is its direct proper function (Millikan 1984, 2000). For instance, indicative-descriptive (fact-stating) syntactic forms in various languages perform a variety of functions (e.g., they impart beliefs, give orders, even deceive), but no form will be labeled "indicative" unless one of its functions is to produce true beliefs (Millikan 2005) having propositional contents carried by other components of the sentence. Indicative syntactic forms continue only because it is in the interests of speakers and hearers that they should. Speakers usually have good reason to share information, for instance, to elicit assistance in some task, and hearers require true information to interpret stimuli and act in the world. The repeated use of false or uninterpretable information by appreciable numbers of people in a speech community would inevitably lead to the extinction of the indicative form.

(ii) Meaning also involves a semantic-mapping function (Millikan 1984, 2005), that is, the mapping (or matching) according to convention of words to states of affairs in the world. It is the application of the semantic mapping function in normal environmental conditions (i.e., naturally recurring local conditions) that accounts for a linguistic device's success in fulfilling its proper function.

It is worth noting at this juncture that the proper function of an expression type in a public language may not accord with the speaker's intention when producing tokens of it on any given occasion. Speakers' intentions in use can lend a language device additional proper functions, as in Gricean implicature, satire, metaphor, and so on. I return to this subject in the next chapter.

(iii) Finally, meaning is a function of the psychological mechanisms involved in the implementation of various linguistic forms, that is, the deployment of conceptions that govern individual speakers' and hearers' grasp of conventional referents and sentential satisfaction conditions.

For language to be satisfactorily coordinated, interlocutors must share a knowledge (but not necessarily explicitly) of both the conventional linguistic functions and semantic mapping functions of their language (otherwise they would become extinct). In contrast, there is no requirement for interlocutors to share or even possess knowledge of the concepts they deploy to track or refer to entities in the world, although it is frequently the case that they do.

For example, we typically perceive sounds, especially speech sounds, as the same sound at origin irrespective of whether we are near the sound origin or far, whether the sound is muffled or clear, whether heard through air or water, and so on. In addition, the perceptual systems use different clues in different situations, separately or together. The perception of distance, for example, is a function of at least ocular disparity, tension in focusing muscles, object occlusion by another, knowledge of size of objects viewed, and atmospheric haze (Millikan 2005). We also recognize distances by touch and stretch using many different parts of the body, and by sound. Importantly, too, we also recognize distances by using ruler and tape measure. None of these methods of distance perception is infallible, but each adds to our concepts of distance. Our sensory-perceptual organs thus enable the construction of concepts of entities we directly encounter "in the flesh" (Millikan 2005, 115). It is the case, however, that we also construct concepts of entities that we do not encounter in the flesh, such as the tooth fairy or Socrates.

Concepts of entities not encountered directly in the flesh are constructed from encounters with language. Language is a standard medium for the transmission of information, and learning to recognize (i.e., perceive) objects and their properties through language is as natural to human beings as through other perceptual channels, such as sight, sound, smell, touch, and taste.

> Recognising a linguistic reference to a substance is as much a way of recognising that substance "in the flesh" as any other way of recognising it. It is identifying it and recognising natural information concerning it through one more medium of manifestation. Think of this medium, the speech of another person, as like an instrument that aids perception. If one wears corrective lenses, these are another such instrument. The speech of another person is analogous to somewhat more complicated instruments of this kind. Like a camera, a radio, a CT scan or a microscope, another person who talks to you picks up information-bearing patterns from his environment, focuses them, translates them into a new medium, and beams them at you.

Think of living in a language community as like being inundated in one more sea of ambient energy. (Millikan 2005, 117)

I noted earlier that we can perceive sound or distance via a range of modalities and in a range of conditions. We can also perceive objects and sounds, however, on surprisingly little information if we are familiar or intimate with them. How much of a parent's or sibling's face do we need to see to recognize them, and how much of "Land of Hope and Glory" do we need to hear to recognize this rugby anthem, at Twickenham, when England is playing at home? This general perceptual ability is a function of a raft of previous and independent experiences of family intimates, favorite anthems, and so on. I could recognize my mother by the look of her face, her body contour, parts of her body, her mannerisms, habits, voice, handwriting, and even traces of her habits (the smell of a Golden Virginia hand-rolled cigarette). These were all natural signs of my mother, which all occasioned the same cognitive response from me, that is, to form another representational token representing some new fact about my mother.

What this implies is that the purpose of perception through language is the same as through any sensory modality, that is, the *tracking* of objects and events of salience to us, not the thinking or reflecting on them (although they may enter sophisticated inferencing processes once these have been developed). Children as young as two years track experiential saliences, including their own and their caregivers' emotions, two years in advance of their acquiring concepts of emotion or of concepts.

6.9 The Mark of the Cognitive

If the fact that an object or process X is coupled to a cognitive agent does not entail that X is part of the agent's cognitive apparatus, what does? The nature of X of course. One needs a theory of what makes a process a cognitive process. One needs a theory of the "mark of the cognitive."

—Adams and Aizawa (2001, 53)

Adams and Aizawa distinguish the mark of the mental from that of the cognitive by noting that qualia and phenomenal consciousness are "not in themselves essential elements of cognitive processes." They take "learning, remembering, sensing, perceiving and thinking as paradigm cases of cognitive processing" (2001, 48). They continue: "A first essential condition on the cognitive is that cognitive states must involve intrinsic non-derived content"; that is, if a process is to be classified as cognitive, it must involve at least some vehicles with nonderived content (2001, 2008, 2010a, 2010b). They do not insist that cognition can only involve the processing of representations with nonderived content (2001), but they do insist that if a process involves no intrinsic content, then the condition rules that the process is noncognitive (2010a, 70).

Nonderived representations include those of trees, rocks, birds, and so on; these images of natural objects acquire their contents by satisfying naturalistic conditions on meaning. They arise "from conditions that do not require the independent or prior existence of either content, representations or intentional agents" (2008, 33). In addition, "whatever is responsible for non-derived representations seems to find a place only in brains" (2001, 63). Derived representations, in contrast, include words, stop signs, warning lights, and so on, and these acquire their meaning by social convention.

Adams and Aizawa's motivation for suggesting that underived content is required is "to explain lone thinkers. Some hundreds of million years ago, the brain of some primitive fish evolved in such a way as to incorporate a fundamentally new type of state. That primitive fish's brain contained a thought or a belief. ... One can speculate about what that first primitive content was" (2008, 32). Adams and Aizawa take pains, especially in their "Defending Nonderived Content" (2005), to show how seriously philosophers take the emergence of the contentful from the content free. They cite Dretske's (1981) informational semantics, Fodor's (1987) asymmetric causal dependency theory, and Cummins's (1996) picture theory, while noting how each is variously limited (Adams and Aizawa 2005, 662–663).

It is worth noting that the emergence of nonderived representations may well explain the first fishy mutation, but even if it does do this, this is all it does. It fails to explain why that initial fishy mutation, rather than proving lethal to the fishy bearer, as gross mutations frequently do, was reproduced in succeeding generations of this particular fish species. A biosemantic theory of mental content does this. In section 6.5, I noted that a biosemantic theory of mental content includes *both* a naturalistic theory of representation *and* a teleological theory and, because of this, includes both biological and social-conventional items. Adams and Aizawa acknowledge only a naturalistic theory in any meaningful sense, and this leads to their focus on the distinction between putatively nonderived and derived representations; it also leads to the implausible implications of their views.

Adams and Aizawa's one-sided view of the mark of the cognitive suggests that the processing of mental representations involves the following:

(i) Those whose contents are nonderived, such as the images of trees, dogs, human faces, and so on. The processing of these images constitutes "Cognition Central" in terms of learning, remembering, sensing, perceiving, and thinking.
(ii) Those whose contents are both nonderived and derived.
(iii) Those whose contents are entirely derived, such as words and numbers. The processing of these does not involve learning, remembering, sensing, perceiving, or thinking.

The implications of these views are clear. To begin with, the off-line reflections of, for example, a theoretical physicist, in terms exclusively of words and numbers, and even the more mundane private soliloquizing of normal human agents, is not thinking. In addition, (Imagistic) Cognition Central is depressingly limited. In terms of human emotional ontogenesis, for example, neonates and children may well form images of human faces, eyes, hands, and so on, but the utility of such images is necessarily restricted to recognition of such features because imagistic representations are nonconceptual. To learn, remember, sense, perceive, and think that these stimuli are faces, eyes, hands, and so on, what they are called in the parent culture, how they function, and so on, and how to use them in human interaction and in doxastic processes, the child requires the addition of derived, conventionally determined representations (and this presumably is why derived representations evolved). Without the addition of conventionally derived meanings, the cognitive capacities of human beings would be limited to those of nonhuman animals.

A biosemantic theory avoids these implausible implications because it does not focus on nonderived representations; rather, it identifies what intentional icons of any ilk have in common. First, they must have had ancestors that conferred a survival or reproduction advantage on their bearers that competitors in the relevant ecological niche did not enjoy. Intentional icons were selected for in ancestral environments, and as such, they were and continue to be produced (copied) by devices that were built by genes (in the appropriate enabling conditions). The constructing genes were and are members of first-order reproductively established families, and the devices that produce (copy) and interpret intentional icons are members of higher-order reproductively established families. Second, and relatedly, intentional icons are produced and interpreted by devices whose proper function is to use the same semantic mapping function (mathematical sense) to bring it about that environmental conditions conduce to agentic well-being. What this implies is that the putative nonderived–derived distinction is redundant. It also implies that if nonderived representations exist at all, they are now as rare as the first fishy thoughts.

A further and related point, as the preceding discussion of human emotional ontogenesis attests, is that the invention of derived representations, in terms particularly of words, has enabled the development of what are humankind's distinctly human capacities. Public language enables the development of our distinctly human learning, remembering, sensing, perceiving, and thinking capacities, which underpin, for example, human emotionality. It is the evolution of derived representations that has made our species distinctly human. Language, it should be acknowledged, evolved from vocal-gestural systems, through formulaic sequences, to fully segmented, generative natural language (Wray 2002), to allow the management of ever-more-complex social relationships and environments (Dunbar 1996). Just as nonhuman animals produce and consume nonlinguistic pushmi-pullyus to manage their worlds, so too do adult human beings produce and consume strings of language devices to manage theirs. Nonlinguistic icons allow nonhuman animals to perceive experiential saliences in their niches and to respond appropriately; nonlinguistically but increasingly as human beings develop and mature (see the discussion of emotional ontogenesis, chap. 4, and this chap., sec. 6.5.3), linguistic icons allow the perception of experiential saliences in their niches and the appropriate responses. The human species developed, as all species develop, typically through evolution by natural selection; the intentional systems with which adult human beings are equipped by evolution and epigenesis, therefore, are as naturalistically acquired as the nonlinguistic intentional systems of nonhuman animals. Derived representations are responsible for our distinctly human cognitive capacities, and these are evolution's most sophisticated achievements. It seems just odd, therefore, to privilege putatively nonderived representations in marking the cognitive.

In addition, Adams and Aizawa (2010a, 73) deny that information processing marks the cognitive; they suggest that it is too general a phenomenon because "DVD players, CD players, MP3 players, tape-recorders," for example, are all information processors (hence the need for nonderived representations to mark the cognitive). They are suggesting, therefore, that the differences between internal and external information processing are such as to preclude functional integration. They conclude: "The weight of empirical evidence supports the view that there are processes that (a) are plausibly construed as cognitive, (b) occur within the brain, (c) do not occur outside the brain, and (d) do not cross the bounds of the brain" (74).

My response to these claims should be obvious; Adams and Aizawa are mistaken, for the following reasons:

(i) The detailed analyses of human emotional ontogenesis included in chapter 4 demonstrate clearly that it is a genuine transcranial phenomenon; the structures and processes on which human emotional ontogenesis supervene quite clearly cross the boundaries of individual brains to include other brains and the sociocultural resources in which they are deeply embedded. Human emotional ontogenesis provides a compelling example of vehicle or locational externalism or transcranialism.

(ii) DVD players, CD players, tape recorders, and so on, may well process information differently and process different kinds of information, and if they do, this is due to differences in their proper function. Such artifacts do possess proper functions; these are the functions their human designers designed them to perform. I have no clear idea of what the proper functions of these devices might be, but it will be the satisfactory performance of these proper functions sufficiently frequently that results in their continued reproduction. In contrast, I do know what the proper function of human intentional devices is; it is for producer and interpreter devices to bring it about (through the operation of the same semantic mapping rule) that environmental conditions conduce to organismic well-being.

Cognitive capacities, that is, learning, remembering, sensing, perceiving, and thinking of differing levels of complexity, enable their bearers to navigate their local environments. Organisms range from the very simple to the highly complex, and the local environments they inhabit range also from the very simple to the highly complex. The cognitive capacities of these organisms evolved to deal successfully with adaptive problems posed by their environments, and indeed, this is reflected in the reactive, enactive, and symbolic representational systems that organisms can differentially possess (R. Wilson 2004). These differing representational systems enable the identification and tracking of, and appropriate response to, environmental objects of salience to the well-being of their bearers. In this account, cognition involves the processing of information of salience to organismic well-being. The identification and tracking of environmental saliences presuppose selective attention to these objects and events, and as I have been at pains to argue, selective attention can be inbuilt (e.g., innate sensory-perceptual constraints) or learned (e.g., emotion elicitors).

The capacities that organisms require to attend to and manage objects and events of salience in their environments differ in complexity in line with the complexities of the local environment they inhabit (this is what "selection for" accomplishes); organisms inhabiting simple environments will need only simple, fixed pushmi-pullyu devices, but those, like the human species, that inhabit enormously complex physical and social environments will need producer and interpreter devices of maximum flexibility. The mechanisms through which human beings acquire such devices, at least in terms of emotion, have been discussed. What is abundantly clear from these analyses is that distinctly human intentional capacities are progressively developed and refined epigenetically, from more primitive precursors, and only in the context of intensive social scaffolding of various kinds. In short, as I have previously argued, there is no cognition without the processing of intentional icons related to environmental salience and organismic well-being, and no distinctly human intentionality without massive social scaffolding involving, overwhelmingly, what Adams and Aizawa term "derived representations."

6.10 Summary

The affect expressions and species-typical behavior patterns of human neonates are the results of brute physiological processes that are interpreted or function as signs of physiological or homeostatic states by preadapted caregivers. This is universally the case and therefore points to a selectionist history. Affect expressions and STBPs have the relational proper function to produce dispositions and then responses in the caregiver to search-identify-correct whatever distresses the neonate, and each particular neonatal emotion producer, once constructed, has the adapted proper function to produce a specific relational structure appropriate to the emotional stimulus. These adapted proper functions have further derived proper functions to produce relational structures of even more specificity (and so on). These relational structures of increasing specificity are produced as a function (mathematical sense) of the semantic mapping rules of human emotionese.

Neonatal affect expressions and intuitive parenting skills are devices that possess the proper function to learn on the basis of experience; they produce behaviors, such as imitation, trial-and-error learning, and beliefs and desires, as a function (mathematical sense) of certain sorts of experiential input. This experiential input is provided mainly by the visual, auditory, tactile, and olfactory sensory-perceptual stimulation provided in mutual neonate-caregiver feedback. The affect expressions and STBPs in the neonate, together with intuitive parenting skills, interact to progressively construct the exquisitely fine-tuned emotional devices in both parties and are the result of mutual adaptation of relational proper function.

Emotional producer and interpreter devices in both parties produce and interpret intentional icons of differential complexity and specificity; these icons include pushmi-pullyus and indicative and directive icons. Neonatal referentially opaque emotion precursors and STBPs and intuitive parenting skills are pushmi-pullyus that are progressively fine-tuned into indicative and imperative icons in response to the mutual feedback typical of neonatal-maternal interaction cycles.

Two conclusions result from these analyses. The first is that transcranialists do not fail to recognize the "mark of the cognitive." Most transcranialists claim that information processing involving neural partial realizers marks the cognitive; my own view, however, although concurring with this more general claim, is rather more precise. It is that the processing of intentional icons of salience to organismic well-being marks the cognitive. The second and related conclusion is that, due to evolution, exclusively intracranial underived representations are illusory. Representations are defined by the way they function, not just intracranially but in extended cognitive systems that include representation-producing devices of differing degrees of complexity and sophistication, objects in the environment, and representation-consuming devices, again of differing degrees of sophistication and complexity. There is nothing in the head, just as such, that displays intentionality.

7 Loose Talk, Tight Worlds

7.1 Introduction

In previous chapters, I have been at pains to argue that the intentional icons that human beings produce, interpret, and use to communicate their emotions and emotion precursors develop from primitive, prelinguistic pushmi-pullyus associated with unfocused affect expressions, through prelinguistic and limited linguistic indicative and imperative intentional icons associated with basic emotions, to complex linguistic icons associated with basic emotions but, more especially, higher cognitive emotions. I have also argued that this developmental trajectory occurs continuously, but only in the context of linguistically mediated social relationships. These relationships are characterized by a range of constraints, ranging from, for example, innate sensory-perceptual competencies to "tight worlds" characterized by locally recurrent local information. The emotional developmental trajectory of human beings is characterized by the interaction of these three Cs (constraints, context, and continuity). This chapter focuses on the linguistic expression of sometimes highly complex emotions, in terms of literal, loose, and metaphorical expression, but even here, constraints continually operate in context to ensure that icon-producing and icon-interpreting devices succeed in communicating the correct meaning of the emotional expression.

Emotional discourse appears typified by apparently surprisingly literal expression (e.g., "you take my breath away") and, rather more frequently, by metaphor (e.g., "you are the sunshine of my life") (Gibbs 2008; Sperber and Wilson 2008; Punter 2007; Knowles and Moon 2006; Stanford 1936). Emotional discourse is replete with metaphor, not only in poetry but also in emotional expression in mundane conversation. This seems to imply that either (a) we require metaphor to express emotions because adequate, literal concepts are unavailable, or (b) aspects of the nature of emotion suit it to metaphorical expression. It also implies, rather obviously, that aspects of the nature of metaphor suit it to emotional expression. In this chapter, I argue that both implications are ill founded.

In previous chapters, I have also argued that both emotional and non-emotional vocal-gestural systems and linguistic strings are ostensive-expressive devices that evolved to attract others' attention and communicate information concerning objects, events, and experiences of salience or relevance to the communication producer, to an interpreter who infers the meaning of the communication (e.g., search-identify-satisfy; see chaps. 4 and 5). In this chapter, and consistent with this ostensive, inferential, and salience-related construal of verbal communication, I apply a relevance-theoretic framework to the explanation of literal, loose, and metaphorical emotional expression. A relevance-theoretic framework construes all intentional communication (in the sense of both "aboutness" and purposiveness) as essentially ostensive and as governed by a principle of relevance. This means that the construal of verbal communication that I adopt is not the usual one of encoding and decoding and governance by maxim, rule, or convention of truthfulness that applies at the level of what is literally meant.

Using a relevance-theoretic framework, I argue that human communication is not a matter typically of encoding and decoding, nor is it governed by maxims or conventions of truthfulness. Human communication is typically a matter of inference, although encoding and decoding do feature as input into inference processes, and it is governed by a principle of relevance, although truthfulness (or some approximation of truthfulness) is one form that relevance can take (i.e., in indicative-declarative intentional icons). I also argue that the same inferential processes operate in the production and interpretation of literal, loose, and metaphorical emotional expression; they differ only insofar as the number and complexity of the implicatures (what the utterance implicitly implicates) and the explicatures (what the utterance explicitly implicates) that they include differ. The theory of emotion and emotional expression that I have developed in this book, therefore, sheds light on the problems associated with figurative expression in the philosophy of language (see sec. 7.2.1).

Metaphorical expression of emotion is commonplace in mundane as well as poetic discourse, and in light of this, I include some rock-bottom preliminaries concerning traditional or orthodox construals of metaphor. I also address the problems perennially associated with metaphor, what Guttenplan (2004, 15) rather interestingly terms metaphor's "three nonnegotiable truths" (3NNTs). These relate to the puzzles concerning the truthfulness, paraphraseability, and transparency of metaphorical expression: metaphor is truthful in some sense (although literally false), it is typically difficult to paraphrase, but its meaning is readily grasped by its hearers. Guttenplan correctly asserts that any adequate explanation of metaphor must explain these puzzles. In light of this, I demonstrate the explanatory adequacy of a relevance-theoretic approach; when using this explanatory framework, the puzzles associated with metaphor evaporate. This is all I claim to do; I do not claim to have developed or even to have begun to develop a general theory of metaphor.

In addition, in previous chapters, I argued that the degree of deep functional integration (DFI) of a range of internal and external factors in emotional and language development is such that they are metaphysically radically widely realized and that the degree of DFI in adult emotional expression is widely realized. In this chapter, I focus on the details of such context dependence in literal, loose, and metaphorical emotional expression. Furthermore, I suggest that, to the extent to which utterance activity and saccadic vision are involved in online language processing, some aspects of language processing are radically widely realized. Finally, I amend Millikan's notion of proper function as it relates to linguistic devices to accord with a relevance-theoretic approach. By way of contextualizing these different analyses, however, I begin with some rock-bottom preliminaries concerning metaphor.

7.2 Metaphor: Some Rock-Bottom Preliminaries and a Very Brief History

The New Shorter Oxford Dictionary defines metaphor as "a figure of speech in which a name or descriptive word or phrase is transferred to an object or an action different from, but analogous to, that to which it is literally applicable" (Brown 1993, 1756). Metaphors figuratively and assertively compare objects, entities, and events to suggest a resemblance and, as such, differ from analogies. Analogies merely infer that two or more entities have similar qualities or properties.

A figure of speech, for example, "Richard is a lion," transfers a name, *lion* (which is termed the source or secondary subject in a metaphorical statement), to an object that is different from, but analogous to, that to which it literally refers (which is the target or primary subject in a metaphorical statement). The metaphor helps people to understand one phenomenon, *Richard*, in terms of its resemblances to another, *lion*. Such resemblances may or may not be readily apparent (Black 1955; Davidson 1978; M. Johnson 1981; Searle 1983). Metaphors therefore point or refer to putative resemblances between objects, entities, and events in the world and their relationships.

Metaphor has had an interesting, if fluctuating, career since the inception of philosophy some two thousand plus years ago in Greece. Depending on the dominant ontological and epistemological construals of the time, metaphor has been seen, on the one hand, merely as stylistic ornament, and on the other hand, playing a central role in epistemology and metaphysics, from being utterly disreputable to being both useful and, indeed, indispensable (M. Johnson 1981).

Metaphor began its philosophical career with Plato. Although he left no fragments with explicit analyses of metaphor, his distrust of poets and sophists and the language that they used so effectively is legendary. His distrust was due to three factors, namely, his view of poets as thrice removed imitators of the Forms, his recognition of metaphor's persuasive power, and his fear of its potential misuse: "Poetry feeds and waters the passions, instead of drying them up; she lets them rule instead of ruling them," and poets "lay on with words and phrases the colours of several arts in such a fashion that other equally ignorant men, who see things only through words, will deem his words most excellent" (*Republic X*, 60la). In short, Plato believed that the use of metaphor and other tropes leads people away from truth (although, even as these short quotations illustrate, he frequently used metaphor himself).

Aristotle recognized, along with Plato, that metaphor was useful in poetry and rhetoric, but his main contribution to the analysis of metaphor lay elsewhere. First, he recognized that metaphor was useful in achieving insight through the perception of similarities in dissimilarities (*Poetics*

1459a). This perception places objects and entities in a new light, enabling them to be seen as they have never been seen before. Altered perspective allows the "get[ting] hold of new ideas" (*Rhetoric* 1410b). Second, he viewed metaphor as an elliptical simile (*Rhetoric* 1406b).[1] This view persists in dictionary definitions of metaphor to this day, even though similes are typically true, whereas metaphors are not (e.g., Juliet is like the sun).

The most common construal of metaphor identification is that when taken literally, an utterance is seen to be defective in some way. Defects include (Beardsley 1962; Searle 1979) patent falsehood; semantic nonsense; violations of the rules of speech acts; violations of the principles of communication, especially Grice's maxim of quality; and lack of congruence with surrounding context, and these despite such violations also characterizing hyperbole and oxymoron (Black 1955). It has also been claimed that the banality of literal readings, in terms of their truth or pointlessness, also points to metaphorical usage (e.g., Black 1955; Guttenplan 2004).

Other ways of identifying metaphor do not relate to word meaning. These include the "announcement" of metaphor when we warn of its imminent use by phrases such as "speaking metaphorically" or, rather more indirectly, "practically," "virtually," and so on. Similarly, Guttenplan (2004, 10) suggests that we identify metaphors simply by recognizing a poetic context in which they occur.

Metaphors are "born," that is, spontaneously generated to impart something of what a speaker wishes to convey; if apt, they are taken up by the relevant speech community and, through use, become conventionalized, that is, used automatically almost as literal expressions; eventually, some, through overuse, become hackneyed and clichéd. Such metaphors are "dead" metaphors and are used as literal expressions.

7.2.1 Metaphor's Three Nonnegotiable Truths (NNT3)

As noted in the chapter introduction, there are three noncontroversial puzzles associated with metaphor; these relate to truth, paraphraseability, and transparency.

(i) Truth

When Romeo states that "Juliet is the sun," his utterance has content with genuine assertoric force with which his audience can sensibly agree or disagree in the ordinary truth-directed way. This is despite his utterance being literally false. It might be discourteous or pointless to tell Romeo that he is mistaken, but in many cases of metaphor, this kind of response is wholly appropriate. What requires explanation is that metaphorical utterances can be challenged or assented to and that their probity or truth is properly regarded as the *responsibility* of their generators (Guttenplan 2004, 15; his italics).

(ii) Paraphrase

Paraphrase quite literally means to say the same thing, to restate the sense of the passage "*in other words*," not just in another way (Guttenplan 2004, 18; his italics). This obviously implies that only words can be paraphrased, but this simple implication can be overlooked. An invitation to paraphrase X is not equivalent to an invitation to state what X tells us or what X means. One can frequently say what an utterance means or tells us without paraphrasing it. In addition, the invitation to paraphrase a metaphor is typically viewed as highly inappropriate. It is rather like asking someone who has just told you a joke to explain it; it kills the joke stone dead. However, when an attempt is made to paraphrase metaphor, this generally is difficult.

(iii) Transparency

Metaphors, even the most complexly poetic ones, are usually composed of words that are familiar to the speech community in which they are used. Because of this, they are usually understood directly and immediately, apparently in the same sort of way that familiar words are understood when used literally (Guttenplan 2004, 21). Speakers of a language just do hear and grasp the meaning of the metaphor immediately. This is not to suggest that the immediate grasping constitutes the kind of deep semantic understanding that might follow sustained reflection on a complex poetic metaphor. It is, rather, merely to suggest understanding in what Guttenplan (2004, 22) terms a "suitably lightweight sense" on first hearing in precisely the same way that we understand undeniably literal utterances when composed of familiar words.

Whatever else a satisfactory theory of metaphor must explain, it has previously been claimed, it must explain NNT3. I will show, in the course of the analyses that follow, that a relevance-theoretic approach to metaphor achieves this goal.

7.2.2 How Metaphors Work

There have been six major views concerning the operation of metaphor since the Greeks first took an interest in it. In rough chronological order, these are the following:

(i) Substitution view (Aristotle): "Metaphor consists in giving a thing a name that belongs to something else" (*Poetics* 1457b). A metaphor of the type "A is B" (e.g., Juliet is the sun) is merely an indirect way of presenting some literal intended meaning "A is C" (e.g., Juliet is radiant).

(ii) Comparison view: A metaphor of the type "A is B" is a means whereby we get the speaker's intended literal meaning, "A is like B in the following respects ..." (e.g., Juliet is like the sun in being radiant, etc.). According to this view, the meaning of the metaphor is a literal set of relevant similarities picked out by the context of the utterance (M. Johnson 1981, 24). This view no longer has adherents; as Black (1955) noted, it suffers from a vagueness that borders on vacuity. The theory fails to describe how the meaning of a metaphor is computed and to explain how the relevant similarities can be picked out.

(iii) Interaction view (Richards 1936; Black 1955, 1993): A metaphor of the type "A is B" involves a system of associated commonplaces of A that interacts with a set of associated commonplaces of B to produce emergent metaphorical meaning. The commonplaces are whatever properties and relations are commonly believed to be true of an object, person, and so on, even if they do not actually apply. For example, "Richard is a gorilla" implies that Richard is fierce even when it is now recognized that gorillas are not typically fierce. Understanding a metaphor, therefore, is not typically a matter of comparing actual properties of objects; it is a matter of what the terms of the metaphor call to mind for us.

(iv) Speech act view (Searle 1979; Cohen 1976): Understanding metaphor is a matter of identifying the speaker's utterance meaning (i.e., what the speaker means by uttering words or sentences with literal meanings) from a word or sentence meaning (i.e., what the word or sentence means literally). Searle identifies three basic steps in understanding metaphor: (i) How does a hearer know how to look for metaphorical meaning in the first place? (ii) What strategies allow a hearer to compute possible values for C where "A is B" is uttered so as to mean "A is C"? (iii) What principles limit the range of possible Cs to grasp the precise meaning of the metaphor? This systematic approach to metaphor comprehension is inconsistent with our everyday experience of metaphor; we typically understand a metaphor immediately (Guttenplan 2004).

(v) Causal theory (Davidson 1978): Metaphors express neither ideas nor thoughts; they belong entirely to the pragmatics of language. The words "Juliet is the sun" mean precisely what they mean literally, that is, Juliet is the sun. Metaphors simply cause associations and reflections as a contingent psychological fact about human speakers. If this were always the case, however, a hearer could never be mistaken in comprehending a metaphor. "Richard is a brick" could just as easily be interpreted as "Richard is a useful doorstop" as "Richard is utterly reliable," which is highly implausible.

(vi) Contemporary view (Lakoff and Johnson 1980; Lakoff 1993; Kovecses 1990): The conceptual systems of human beings are, typically, fundamentally metaphorical in nature. Metaphor is pervasive in language, thought, and action. Conceptual systems are central in defining everyday realities; the way we think, what we experience, and what we do are very much a matter of metaphor. "Argument is war," for example, is not merely a linguistic expression; we actually do win or lose arguments. Our interlocutor is viewed as an opponent, we attack his position, we defend our own, we gain and lose ground, and so on. We live by the "argument is war" metaphor in our culture; it structures the actions we perform in arguing (Lakoff and Johnson 1980).

Each view includes a construal of the mechanism of metaphor and the level at which it operates, that is, at word, sentence, or some deeper level. In addition, each view (contemporary view excepted) is founded on a construal of human language as code; I will not, therefore, discuss them further here.

7.3 The Code Model of Communication

That metaphor is typically considered a distinctive linguistic device is a function of the generally considered construal of language use as governed by a norm of literalness that follows naturally from another widely held construal, namely, that of the function of language in communication. The function of language is to allow the speaker to encode the meaning of her thoughts, and the listener to decode such meanings; this construal is beautifully exemplified in Reddy's (1979/1993) metaphor of language as conduit. It implies that (i) the interpretation of metaphor follows a failed literal interpretation; (ii) because of this, processing time is increased; and (iii) context is unimportant to metaphor interpretation. All three implications are mistaken (see hereafter); the code model is therefore untenable.

A code is a systematic pairing of messages with signals and is extremely common in nonhuman animal communication. To communicate entirely by code, however, each signal in the code must unambiguously convey exactly the same message on all occasions of use. The vervet's alarm call "leopard! here! now!" to her group of conspecifics is exemplary in this respect. Ambiguity, where the same signal is paired with several messages, simply prevents the decoding process. Of course, there are examples in the nonhuman world where the exact message encoded by a given signal varies with context; the dance of the honeybee is an example of this. A coding–decoding system can accommodate such limited context sensitivity but cannot support the overwhelming complexity of human languages. Human languages are codes in the same sense that they are pairs of phonetic-semantic representations generated by an underlying grammar (Sperber and Wilson 1995, 9), but compared to nonhuman animal codes, they are incomparably richer (Sperber and Wilson 2008, 85). Human languages contain a vast repertoire of expressive elements (the lexicon), and these elements are combined by a syntax with unlimited generative capacities. In addition, the sentences of natural language are multiply ambiguous; they contain referential expressions whose values cannot be assigned by decoding alone, and the senses they encode are often elliptical or incomplete. This implies that a gap exists between the semantic structure a sentence encodes and the meaning a speaker manages to convey by uttering it in a given situation. This in turn implies that the linguistic content of all sentences, including those that are literally understood, vastly underdetermines their interpretation (Feldman 2006; D. Wilson and Sperber 2002; Sperber and Wilson 1986, 1995).

This underdetermination reflects important differences between sentences and utterances. An utterance has a variety of properties, both linguistic and nonlinguistic. It may contain the word *dog* or a reflexive pronoun or a trisyllabic adjective; it may be spoken clearly or by someone with a heavy cold, loudly or whispered, in a supermarket or a bedroom. Generative grammars abstract out the purely linguistic properties of utterances and describe a common linguistic structure, the sentence, shared by a variety of utterances that differ only in their nonlinguistic properties. By definition, a semantic representation of a sentence, as assigned to it by a generative grammar, can take no account of nonlinguistic properties such as time and place of utterance, identity of speakers, and speakers' intentions (Noveck and Sperber 2004).

The semantic representation of a sentence deals with a sort of common core of meaning shared by every utterance of it. However, different utterances of the same sentence may differ in their interpretation, and they typically do:

(1) I am sad.

(2) Bill is angry.

(3) Betty's gift made her happy.

Generative grammar cannot pick out who "I" am, which Bill "Bill" picks out; it can only provide some general indications. "I" always refers to the speaker, and "Bill" and "Betty" belong to people with those names. None of this, however, is sufficient to determine the thought expressed in sentences (1) to (3). There is no mention of how sad I am or how angry Bill is, and what sense of "gift" makes Betty happy. These aspects of interpretation involve interaction between linguistic structure and nonlinguistic information, only the structure being dealt with by the grammar (Sperber and Wilson 1995, 10).

Further facts widen the gap between sentence meaning and utterance interpretation. One example is that one sentence, used to express the same thought, may sometimes be used to present the information as true, sometimes to suggest that it is not, sometimes to wonder whether it is true, and so on. Utterances are used not only to convey thoughts but to reveal the speaker's attitude or relation to the thought expressed. Utterances express propositional attitudes, perform speech acts, or carry illocutionary force (Sperber and Wilson 1995, 11).

Another example is when an utterance explicitly expresses one thought but implicitly conveys another. Whereas a thought that is explicitly expressed must be in some correspondence to the semantic representation of the sentence uttered, those that are implicitly conveyed are not so constrained:

(4) *Jack*: Do you want to go to the movies on Thursday?

(5) *Jill*: Jane is coming to dinner.

Another source of underdetermination may be that the literal interpretation need not be the one intended and may not be the preferred interpretation, for example, "pigeonhole," "seed funding," "jump down his throat," "move heaven and earth" (Knowles and Moon 2006, 12, 21).

This is not to suggest that human languages are grossly defective; on the contrary, that they survive, flourish, and evolve testifies to their excellence in fulfilling their function of communication. It is to suggest, rather, that the function of language cannot be simply to encode speakers' meanings. In human communication, coding is surprisingly limited.

As emphasized throughout the foregoing chapters, human beings communicate not only with language but also by producing other nonlinguistic ostensive stimuli that are deliberately intended to attract the addressee's attention and convey some content (e.g., unfocused crying, motor unrest). These ostensive stimuli include gestures, such as pointing and writing, as well as speech. Some of these are not codes, and yet human beings, even very young ones (see chaps. 4 and 5), can communicate extremely successfully without language at all.

It is worth emphasizing at this juncture that human communication is inferential communication. The communicator provides some evidence of her meaning in what she says and what she does, and her addressee infers the meaning on the basis of this evidence and the context. Linguistically encoded utterance is just another source of evidence, but it provides evidential resources that can be as "nuanced, as complex, as richly structured as the speaker likes. ... With language, and only with language, people can communicate about anything they can think about, whether they can point to it or not, imitate it or not, and they can do this with almost limitless refinement" (Sperber and Wilson 2008, 87). Language thus enables the expansion of emotional expressive repertoires when the emotions become more subtle, nuanced, and refined, as demonstrated by my previous discussion in chapter 5 of language acquisition in infants and children.

7.4 Relevance Theory: A Brief Introduction

The theories of David Lewis (1975/1996) and, more particularly, Grice (1989) resulted from a recognition of sentence and utterance distinctions: they both reflect an inferential model of human communication. They tend to assume, however, that the inference from sentence meaning to speaker meaning is a function merely of assigning referents to referring expressions and calculating implicatures. Maxims or conventions of truthfulness reduce the gap to a minimum. This is despite the enormous complexity of the comprehension process, which involves both decoding and inferential processes. A relevance-oriented inferential framework allows for much greater slack between sentence meaning and speaker meaning than has hitherto been assumed. It also provides a general framework adequate to the interpretation of literal, loose, and metaphorical expression.

Relevance has been characterized as a property of inputs to cognitive processes that makes them worth processing (D. Wilson and Sperber 2002, 601). Such inputs may be external stimuli (e.g., a smell, a loud noise) or internal representations that may undergo further processing (e.g., the recognition of a smell, the memory of a loud noise, the linguistic decoding of an utterance). Of course, each human agent is potentially bombarded at any given moment with many more stimuli than she can process, either from the environment or from memory (we have many more memories than we can activate at any given moment). Cognitive efficiency, therefore, is a result largely of allocating processing resources to maximize cognitive benefit. This involves processing the inputs that offer the best expected cost-benefit value at the time (601).

An illustration: Peter wakes, feeling unwell, and goes to his doctor. She examines him carefully and could say any of the following:

(6) You are ill.

(7) You have flu.

(8) You have flu or 29 is the square root of 841 (D. Wilson and Sperber 2002, 602).

Noticeably, the literal content of all three utterances is relevant to Peter. Clearly, however, (7) is more relevant than either (6) or (8). It would be more relevant than (6) in terms of cognitive effects; it yields all the information of (6) plus additional information. Utterance (7) is more relevant than

(8) in terms of processing effort; they both yield the same consequences, but the consequences are derived more easily from (7) than (8), which requires an additional effort of parsing and inference (to realize that the second disjunct is false and the first is therefore true).

The tendency to process the most relevant inputs available is so obvious to us that human beings can render their mental processes at least partially predictable to others. We are capable of identifying which external stimuli another person is likely to attend to and accurately predict what conclusion, from the many available, she will in fact draw. We assume that she will attend to the potentially most relevant stimulus and process it to maximize relevance, that is, in the context of easily accessible background assumptions where the information it provides will carry relatively rich cognitive effects (D. Wilson and Sperber 2002, 603). This mutual predictability is exploited in communication.

As communicators, we provide stimuli that are likely to strike our intended audience as relevant enough to be worth processing, and to be interpreted in the intended way. A communicator produces a stimulus, such as an utterance, that attracts her audience's attention, and she does so in an overtly intentional way. In other words, she makes it manifest that she wants the audience's attention. Since it is also manifest that the audience will tend to pay appropriate attention only to an utterance that seems relevant enough, it is manifest that the communicator wants her audience to assume that the utterance is indeed relevant enough (D. Wilson and Sperber 2002, 604).

What this implies is that listeners are encouraged to expect a minimum level of relevance. The utterance should be sufficiently relevant to warrant the listener's effort of comprehension. However, an utterance is most likely to be comprehended when it simplifies the listener's task by demanding as little processing effort as possible but encouraging her to pay due attention by offering as much effect as possible. "The smaller the effort, the greater the effect, the greater the relevance" (D. Wilson and Sperber 2002, 604). What this in turn implies is that it is manifestly in the speaker's interest for the hearer to presume that the utterance is more relevant than merely sufficiently relevant to be worth her attention. The implication of this, again in turn, is that the speaker should provide the most appropriate information in terms of quality and quantity and present it in the most efficient way. Both of these factors—that is, what is included in the utterance and how it is presented—are a function of the speaker's own communicative abilities, what she hopes to achieve in communicating, what cultural norms or rules of etiquette she wishes to observe, and so on, and the interpretive abilities she expects the listener to possess. When a speaker considers all these factors, she is aiming at an optimum level of relevance (of an utterance, for a listener). Thus, for example, when talking to her infant, a caregiver will use baby talk to comment on what she is doing and what the infant is doing, often using two-word sequences of just noun and verb.

As previously noted, when interpreting an utterance, there are indeterminacies, including illocutionary indeterminacies, ambiguities, and other indeterminate aspects of explicit content and implicatures to resolve, ironies and metaphor to interpret. All of this requires an appropriate set of contextual assumptions. Given that the speaker is expected, within the limits of her abilities and preferences, to ensure that her utterance is "of optimal comprehensibility, the hearer should consider interpretive hypotheses (disambiguations, reference assignment, implicatures, etc.) in order of accessibility. In short, the hearer should follow a path of least effort and stop when she arrives at an interpretation which satisfies the expectation of relevance raised by the utterance" (D. Wilson and Sperber 2002, 605).

Relevance varies inversely with processing effort; the very accessibility of the first hypothesis renders it, or should render it, the correct one. This reflects a *relevance-guided comprehension heuristic*:

(i) Follow a path of least effort in constructing an interpretation of the utterance (and, in particular, in resolving ambiguities and referential indeterminacies, going beyond linguistic meaning, supplying contextual assumptions, computing implicatures, etc.).
(ii) Stop when your expectations of relevance are satisfied.

A hearer using the relevance-theoretic comprehension heuristic during online comprehension should find an interpretation of speaker meaning that satisfies the presumption of optimal relevance. To achieve this, she must do the following:

(iii) Enrich the decoded sentence meaning at the explicit level and complement it at the implicit level by supplying contextual assumptions that will combine with it to yield enough conclusions (or other cognitive effects) to make the utterance relevant in the expected way.

(iv) Follow a path of least effort and stop at the first overall interpretation that satisfies her expectations of relevance.

7.5 Explicatures and Implicatures

Explicatures refer to the explicit contents of utterances, and implicatures to their implicit contents. Explicatures are identified through a combination of decoding and inference, while implicatures are wholly inferred:

(9) *Lisa*: What is the matter with you?

(10) *Alan*: You take my breath away.

Alan's reply might include the explicature that he has just been robbed of breath by unexpectedly seeing Lisa and the implicature that he has been rendered breathless because he finds her beautiful. The greater the relative contribution of decoding and the smaller the relative contribution of pragmatic inference, the more explicit an explicature will be and vice versa (Sperber and Wilson 1995, 182). The weaker the explicature, the harder it is to paraphrase what the speaker was saying.

When the speaker's meaning is quite explicit as in (10), and, in particular, when each word in an utterance is used to convey one of its encoded meanings, the explicature is close to what might commonsensically be termed explicit content, or what is said, or the literal meaning of the utterance. The less explicit the meaning, the more responsibility the hearer must assume in interpreting the utterance.

The recovery of both explicit and implicit content may involve a substantial element of pragmatic inference. The linguistically encoded meaning of an utterance gives no more than a schematic indication of the speaker's meaning. The hearer's task is to use this indication, together with background knowledge, to construct an interpretation of speaker meaning, guided by expectations of relevance raised by the utterance itself. This requires that the hearer must do the following:

(1) Construct an appropriate hypothesis about explicatures by developing the linguistically encoded semantic schema.
(2) Construct an appropriate hypothesis about the intended contextual assumptions (implicated premises).
(3) Construct an appropriate hypothesis about the intended contextual implications (implicated conclusions).

These subtasks are undertaken simultaneously. The hearer does not first decode the sentence meaning, then construct an explicature and identify an appropriate context, then derive a range of implicated conclusions. Comprehension is an online process, and hypotheses about explicatures, implicated premises, and implicated conclusions are developed in parallel, against a background of expectations that may be revised or elaborated as the utterance unfolds. In particular, the hearer may bring to the comprehension process not only general assumptions of relevance but more specific expectations about how the utterance will be relevant to her (what implicated conclusions she is intended to derive), and these may contribute, via backward inference, to the identification of explicatures and implicated premises. The overall process is guided by the relevance-theoretic comprehension heuristic outlined earlier. In this approach, explicatures and implicatures are constructed by mutually adjusting tentative hypotheses about explicatures, implicated premises, and implicated conclusions to satisfy the expectations of relevance raised by the utterance itself.

Implicatures, like explicatures, may be stronger or weaker, depending on the degree of indeterminacy introduced by the inferential element of comprehension. When the hearer's expectations of relevance can be satisfied by deriving any one of a range of roughly similar conclusions, at roughly comparable cost, from a range of roughly similar premises, the hearer also takes some responsibility for the particular premises he supplies and the conclusions he derives from them.

7.6 Loose Talk

Loose talk includes rough approximations and sense extensions where an expression is applied to items that fall outside its linguistically determined denotation, strictly understood. The decoded senses of a word or other linguistic expression in an utterance are the starting point of an inferential process of meaning construction. The meaning constructed can be narrower than the decoded meaning or broader:

(11) I have a temperature.

Here temperature would be understood as narrowed to "temperature above normal." If the speaker had a normal temperature (37°C), then, strictly speaking, her utterance would be false. This is because every living creature, the air, the sea, and so on, all possess the property of temperature, and to state "I have a temperature" when one's temperature is normal is simply irrelevant. As this example demonstrates, in narrowing, some aspect of literalness is preserved; a high temperature is still a temperature. This distinguishes narrowing from broadening:

(12) I have a cold. I need a Kleenex.

This is an example of category extension or broadening. Category extension involves extending a word with a relatively precise sense to a range of items that clearly fall outside its linguistically specified denotation but share some contextually relevant properties with items inside that denotation. What the speaker needs is a paper tissue, any paper tissue with which to wipe her nose. Thus (12), strictly speaking, is false.

It is important to recognize, however, that although the preservation of some literalness in narrowing and broadening differ, they are not functionally two distinct types of language use. They both involve the same process of meaning construction that leads, in some cases, to a broadening of the encoded concept and, in others, to a narrowing of it.

These narrowed or broadened lexical meanings are constructed by following the relevance-guided comprehension process outlined earlier. As already indicated, when (11) is interpreted literally ("I have a temperature"), it is simply an irrelevant truism that the hearer neither constructs nor entertains. What happens is that the concept TEMPERATURE is activated in the hearer's mind and points him toward a relevant interpretation. This concept has a range of values, some of which would be relevant in the circumstances (by implying, e.g., that the speaker is unwell). As the hearer constructs a relevant overall interpretation of the utterance, the decoded concept TEMPERATURE provides a starting point for constructing a narrowed ad hoc concept TEMPERATURE*, which ranges over only contextually relevant temperatures, that is, temperatures that are abnormally high. These are easily brought to mind, with implications that they are worth the hearer's processing effort.

The word *temperature* in this example is used in an utterance to activate potential implications of the encoded concept TEMPERATURE. Ideas activated in utterance interpretation stand in inferential relation to the concepts that activate them and are not mere associations based on past co-occurrence with no inferential status (Sperber and Wilson 2008, 92). The ideas activated by a word in an utterance are likely to be true of items in the linguistically specified denotation of the word or, equivalently, of items in the extension of the concept encoded by the word. In narrowing, as in "I have a temperature," the implications involve only part of the extension of the encoded concept (because only some temperatures imply malaise). In broadening, the implications involve not only items in the extension of the encoded concept but also contextually salient items that fall outside the extension but share with items inside the extension properties that determine these implications.

Of course, implications are not just activated by words in utterances; they are frequently activated by events or objects in the context of the utterance. I would offer someone a Kleenex if I had just witnessed her sneezing (Sperber and Wilson 2008, 92). And "Here is a Kleenex," said of any paper tissue, activates the implication in the hearer's mind that it can be used to blow her nose when she has just sneezed. Clearly, implications that are activated both by words in utterance and by events and objects in the conversational context are the first to come to mind and are tentatively added to the interpretation until the hearer's expectations of relevance are satisfied. This is how caregivers infer the meaning of their toddlers' loose talk, which begins at around two to three years of age. If the child does not yet possess the word for what it is she wants to convey, she selects one that represents similar features. Parents have no difficulty inferring from the utterance context and the child's gestural accompaniments what she is trying to say. What this implies, rather obviously, is that loose talk is typical in human speech from the beginning of its ontogenetic acquisition.

Once expectations of relevance are satisfied, the explicit content of the utterance (in the case of assertion, the proposition whose truth the speaker is committing herself to) is retroactively determined by mutually adjusting the implicit and explicit components of the interpretation. The explicit content must be such that it contextually implies the implicit content. More technically, the explicatures of an utterance must be such that, together with the implicit premises of the utterance, they justify the derivation of its implicit conclusions (Sperber and Wilson 1998, 2005, 2008; D. Wilson and Sperber 2002, 2004).

In the case of "I have a temperature," the result of the mutual adjustment process is a contextual construal of "temperature" as TEMPERATURE*, which is narrower than the lexicalized concept TEMPERATURE. Strictly literal interpretations, that is, those that involve neither narrowing nor broadening of the lexicalized concept, are arrived at by the same mutual adjustment of implicit and explicit content. A literal interpretation results when the implications that make an utterance relevant in the expected way depend on the explicit content of the lexicalized concept itself (which has undergone neither broadening nor narrowing). What this implies is that literal interpretations are not default interpretations; they are not the first to be considered, nor are they necessarily easier to construct than nonliteral ones.

It is worth noting, even at this juncture, that metaphors generally but not invariably involve both narrowing and broadening:

(13) You are the sunshine of my life.

Here "sunshine" is narrowed to exclude all the negative effects of sunshine (e.g., burning, glare) and broadened to include all beloved people (Sperber and Wilson 2008).

Loose use of language also includes rough approximations:

(14) The lecture starts at five o'clock.

When interpreted in the most restrictive way (ignoring issues of ambiguity and polysemy), this utterance is not strictly true; lectures very rarely start at five o'clock, precisely. For instance, you have a philosophy lecture this afternoon but do not know what time it starts. You are told "five o'clock." From the utterance of this phrase, you may draw a number of inferences that make the utterance relevant to you:

You cannot go shopping at five o'clock.
You should leave the library by 4:50 p.m.

Yet neither of these depends on a strict interpretation of five o'clock. According to the relevance-theoretic approach, you take the speaker to be committed to the truth of the proposition that warrants just the implications you derive, a proposition that might be paraphrased as "between 4:50 and 5:10," which you, as hearer, have no need to formulate mentally. And, importantly, had the speaker uttered the strictly true longer phrase instead of the shorter approximation, the processing effort would have been increased but without additional cognitive effects.

Such loose uses are extremely common. Some may be tied to an idiosyncratic situation, used once, and then forgotten. Others, such as Kleenex, may be used frequently enough to give rise to an extra sense that may stabilize in an individual or population; lexical broadening (e.g., Kleenex means any paper tissue), along with lexical narrowing and metaphorical transfer, is one of the main pragmatic factors driving semantic change (Sperber and Wilson 2008).

The point is that if verbal communication was driven by expectations of truthfulness, such loose talk would be viewed as the result of sloppy thought or speech, accepted by hearers "whose expectations have been reduced to realistic levels by repeated encounters with normal human failings" (D. Wilson and Sperber 2002, 592). As hearers and speakers, we should always prefer literally true utterances like the following:

(15) The lecture commences around five o'clock.

(16) I need a Kleenex or other disposable paper tissue.

Yet clearly we do not. Hearers are not misled by the previous uses of (12) and (14), and their strictly true versions provide no additional valuable information. An additional point is that (15) and (16) are appreciably longer utterances than (12) and (14) and require extra processing effort. In cases like these, (12) and (14) are seen as preferable. Speakers and hearers are so used to loose talk that they generally fail to notice its occurrence.

The use of hyperbole, litotes, irony, synecdoche, metonymy, and metaphor is also illustrative of the limitations of an expectation of truthfulness in utterance; none are literally true. These tropes and figurative devices point to a range of potential interests operative in human cognizing rather than just one. Truthfulness is an expectation in indicative-declarative sentences; indeed, as previous biosemantic analyses demonstrate, truth is a necessary prerequisite to their continuing use. When someone wishes to know what is the case, then truthfulness is expected, but only because and to the extent that it is relevant to her aims. However, when a speaker wishes to convey more than the linguistic meaning of an expression such as "your room is dirty and untidy," she can say:

(17) You are a piglet (metaphorical expression).

(18) You are a pig (another metaphorical expression) (Sperber and Wilson 1986, 153–154).

(19) Did a tsunami pass through here? (irony).

Each one of these conveys something in addition to "your room is dirty and untidy": (17) conveys additional implicit information concerning the endearing nature of the addressee, which is absent from (18). Piglets, as baby animals, are generally endearing, whereas pigs, as adult animals, generally are not. In fact, to call someone a pig is (or can be) seen as insulting. Utterance (19), as irony, could point to the degree of untidiness in the room, since the effects of tsunami are usually devastating. And again, the use of irony can serve either to soften or to increase the degree of rebuke intended. All of these are relevant to the differing interests an utterance can be meant to convey.

7.7 Cognitive Environment

All humans live in the same physical world. Agents are all engaged in deriving information from this environment and constructing the best possible representations of it. Agents inhabit smaller locales than the world itself, however, and are differentially endowed in terms of perceptual, conceptual, and inferential abilities. In addition, different cultures speak different languages and include different concepts. This means that the representations people construct and store in memory differ. What this implies is that people possess and inhabit different cognitive environments (Sperber and Wilson 1995, 38). An individual's cognitive environment is the set of facts that are manifest to her, *manifest* meaning that she is capable at that time of representing the fact and accepting it as true or probably true (39). To be manifest, in this account, is to be perceptible or inferable. An individual's total cognitive environment is a function of both her physical and social environment and her cognitive abilities, including memory.

Cognitive environments, however, include not only facts; they include assumptions that may or may not be true but result from, and enter into, inference processes as if they were true. An additional point is that manifestness is a comparative concept; facts and assumptions can be more or less manifest. The stronger an assumption is manifested, the likelier it is to be entertained. Which facts and assumptions are manifest to an individual at any given time is a function of her physical and social environment and cognitive abilities.

Human cognition, it has been noted, is innately equipped with certain salience detectors. Certain phenomena (loud noises, large objects) are taken as highly salient, whereas less-noisy sounds and smaller objects are not. For instance, a doorbell ringing is highly salient, but the ticking of a clock is not. When a phenomenon is noticed, some assumptions about it are standardly more accessible than others. In an environment where the doorbell has just rung, the hearer will normally strongly assume that someone is at the door, less strongly assume that whoever rang the bell is tall enough to reach it, and even less strongly assume that the bell has not been stolen (Sperber and Wilson 1995, 40).

Another point worthy of attention here is that two people who share a physical environment are mutually aware that they share it; it is mutually manifest that they share it. If Jack and Jill are in the kitchen when the doorbell rings, they will both hear it and be aware that they both hear it ("Will you get that?"). This mutual cognitive environment provides the background in which coordinated choice of code and context becomes possible. A mutual cognitive environment is strong enough to enable the speaker to correctly ascertain what codes and contextual information the hearer has access to so as to comprehend the utterance.

From assumptions about what is manifest to other people and, in particular, about what is strongly manifest to them, we are in a position to derive further, though necessarily weaker, assumptions about what assumptions they are actually making. From assumptions about what is mutually manifest to all of us, we are in a position to make assumptions about the assumptions they attribute to us.

These analyses are correct as far as they go, but they fail to go far enough. They fail to emphasize sufficiently how much of our mundane experiences are the same, how much they are repeatedly perceived, day in, day out. It is this sameness of locally recurrent signs of locally recurrent information (see chap. 6) that renders efficient communication possible.

To begin with, the experiences of human neonates are tightly constrained by their immobility and underdeveloped perceptual capacities. Their worlds are limited to cyclical rounds of sleeping, feeding, changing, and interacting or playing with a limited range of caregivers, at least initially. They sleep in the same cot, in the same bedroom, with the same wall decorations, mobiles, and so on. The faces of their caregivers are the same, even if their expressions change. Their caregivers not only look the same but smell the same and sound the same. As the child grows and her perceptive cognitive repertoire develops and expands, so too does her activity range. But the sameness continues. Friends are recognized only because they remain the same; the furniture in the home, equipment in the gym, the playground, the rules of sporting and other games remain the same. Weather is cyclical, together with flora and available food. In addition, from the outset, the child learns the language and concepts of her culture from her caregivers. Recognition is only possible (obviously) because objects, people, and events remain more or less the same (locally recurrent signs of locally recurrent information).

7.8 Metaphor's 3NNTs

It should be clear from the foregoing discussion that a relevance-theoretic approach to communication satisfies Guttenplan's desiderata, for the broadening and narrowing of the extension of the lexicalized concept, the target of the metaphor, is responsible for metaphor's apparent truthfulness, lack of adequate paraphraseability, and transparency. In broadening, the implicatures involve not only items that fall within the extension of the lexicalized concept but also contextually salient items that, although falling outside its normal extension, share certain properties with items within the extension that determine such implicatures. In narrowing, although the ad hoc concept clearly falls outside its linguistically specified denotation, it shares some contextually relevant properties that fall within that denotation. This accounts for metaphor's apparent truthfulness and transparency. And it is clear why it accounts for metaphor's relative lack of paraphraseability. Metaphor does not use normal lexicalized concepts directly; it merely uses them to indicate some features of the items they represent that resemble their target items. Their relative lack of paraphraseability is a function of the number and complexity of the implicatures they include.

7.9 Conclusions

The absence of any criterion for distinguishing literal, loose, and metaphorical utterances suggests two things: first, there are no genuinely distinct categories as literal, loose, and metaphorical, at least not from a descriptive or pragmatic perspective; second, there is only one inferential procedure for interpreting all three types of utterance. If metaphorical interpretation was, in fact, dependent on a prior attempt at literal interpretation, processing time would inevitably be increased (for what amounts to a two-stage process). This is not the case. The processing of metaphor takes no longer than the processing of putatively literal statements (Gibbs 2008). An example of this common procedure for both literal and metaphorical utterances is provided hereafter (adapted from Sperber and Wilson 2008, 95–96).

(a) Inferential Procedure in a Literal Utterance

(i) Lover says to beloved, "You take my breath away," uttered in response to interoceptive subjective feelings of pleasant breathlessness (beloved's decoding of lover's utterance).

(ii) Lover's utterance is optimally relevant to beloved (expectation raised by the recognition that lover's utterance is a communicative act, i.e., is an ostensive stimulus).

(iii) Lover's utterance will achieve relevance by beloved observing his breathless state, slightly gasping speech, and enraptured expression.

(iv) Since the couple are face-to-face (and probably have just come face-to-face) and the beloved knows from experience that the sudden sight of something beautiful can take one's breath away, the enraptured look eliminates other possible causes of breathlessness, such as vigorous exercise or fright (beloved's tentative acceptance of this as implicit premises).

(v) The beloved is beautiful in the eyes of her lover (beloved's implicit conclusion derived from [iii] and [iv] and accepted as interpretation of lover's utterance by relevance-guided comprehension heuristic).

(b) Inferential Procedure in Metaphorical Utterance

(i) Lover says to beloved, "You are the sunshine of my life" (beloved's decoding of lover's utterance).

(ii) Lover's utterance is optimally relevant to beloved (expectation raised by recognition of lover's utterance as communicative act, i.e., as an ostensive stimulus).

(iii) Lover's utterance will achieve relevance by beloved's observing the enraptured face and soft verbal tones of the lover (expectation raised by [ii], given that the lover is looking at his beloved).

(iv) Sunshine brightens and lights up the world, intensifies color, is generally pleasant to experience, is warm, encourages plant growth, encourages outdoor recreation, and so on (assumptions activated by word *sunshine* and lover gazing enraptured at beloved's face. Some or all of these are accepted as implicit premises of lover's utterance).

(v) The beloved is a sunshiny person; she lights up her lover's life, intensifies his experiences, makes him feel warm, and so on (beloved's implicit conclusion derived from [iii] and [iv] and accepted as interpretation of lover's utterance, by relevance-guided comprehension heuristic).

In both literal and metaphorical interpretations, the interpretation is carried out online and commences while the utterance is still in progress. This implies that the interpretive hypotheses concerning both explicit content and implicatures are developed at least partly in parallel rather than in sequence. They stabilize when their mutual adjustment confirms the audience's expectations of relevance.

Of course, hearers do not consciously proceed through this inferential process or, if they do, do so only until they understand the meaning of the lexicalized items used. Meaning perception qua meaning perception, even when assisted or augmented by language, generally goes on automatically and effortlessly once the perceiver has learned to identify the object or event.

In terms of biosemantics, it is clear that Millikan's analyses relating to the direct and derived proper functions of linguistic devices, including declarative-indicative and directive-imperative icons, are underpinned by a code theory of communication. She construes speech clearly as a form of direct perception of whatever the speech is about (see chap. 6); interpreting speech does not require making any inferences or having any beliefs about the speaker's intentions. It is a sufficient condition for linguistic communication that the linguistic devices used succeed in performing their proper function. Indicative utterances, it has been noted, continue to be produced because they produce beliefs in hearers, and imperative utterances produce compliance. Yet Millikan does concede that understanding a language is never just decoding (1998, 176); further processes must use the output of decoding and information about the situation to fix the contextual meaning of the utterance. These further processes, however, consist merely in strict disambiguation, that is, the selection of one of the possible decodings of the utterance. All the possible contextual meanings of a linguistic device must be conventionally associated with the device. This implies quite extraordinary ambiguity of nearly all linguistic utterances (Origgi and Sperber 2000). According to Millikan, the hearer's task is to recognize to which one of the families that proliferate phonetically or syntactically indistinguishable but semantically different tokens that this particular token belongs.

Yet almost infinitely many new meanings (generated syntactically and compositionally) can be communicated by the same means of the same linguistic device when used perfectly normally, for example, "go home." This means that disambiguation (identifying from which family these linguistic tokens were reproduced) will not explain the contextual aspects of comprehension. "Meanings are not just disambiguated, they are in part disambiguated, in part constructed in context" (Origgi and Sperber 2000, 152).

Millikan is correct in claiming that the direct proper function of a device is that which keeps speakers and hearers reliably producing and responding to the device, thus stabilizing the device in a language community. This response, for Millikan, is to be found at the level of belief or desire formation or even at the behavioral level of compliance. The function of a word is to convey "conventional meaning" to the overall meaning of an utterance, which will be accepted as belief or desire by the hearer. The same linguistic device, however, may elicit a wide range of responses at the belief or desire level. It has been argued that language comprehension involves the decoding of linguistic stimuli that is then used by the hearer, together with contextual information, to infer the speaker's meaning. The same unambiguous linguistic device, decoded in the same way each time it is produced, can serve as evidence of a range of different meanings in different contexts ("John is a computer"). The direct proper function of a linguistic device, therefore, is to provide reliable evidence of the speaker's meaning. Consistent with this, the derived proper function of a linguistic device is to reliably provide evidence of the speaker's meaning in a particular context. In short, the direct proper function of, for example, the word *piglet* is to give evidence that the speaker's meaning includes a concept, PIGLET, which may be, but may not be, the concept lexically encoded by *piglet*. This direct proper function is performed through each token of *piglet* performing the derived proper function of evoking, in context, a specific concept that is part of the speaker's meaning on that occasion (Origgi and Sperber 2000, 159).

All online language processing, in terms of both production and comprehension, is at least widely metaphysically realized. A raft of contextual factors, such as knowledge of each other's cognitive environment and visual access to shared physical environment, are constitutive of meaning construction for both speaker and hearer. The more numerous and weak the implicatures in any utterance, the more production and comprehension are context dependent.

How much of online processing is radically widely realized is an interesting question. Utterance activity, as Spurrett and Cowley (2010) and Cowley (1998) construe it, always underpins and accompanies linguistic expression, and so too does saccadic vision. At least elements of utterance activity are coregulated, that is, mutually and synchronously modulated, and saccadic vision provides a straightforward example of locational or vehicle externalism. What this implies is that whenever elements of utterance activity or saccadic vision are involved in utterance comprehension (and it is difficult to imagine when they might not be), the core realizer system extends beyond the skin and skull of both interlocutors.

8 Once More, with Feeling

8.1 Introduction

In this chapter, I sum up the most significant theoretical and conceptual achievements cataloged previously and discuss their ramifications for philosophical thinking (and thinking in other disciplines) in terms of human emotional and language development and for empirical research relating to these. I sum up research results that clearly suggest that an internalist, individualist, and intracranialist metaphysics of mind should be abandoned; such a metaphysics fuels, if not actually generates, the range of putative dichotomies that this monograph explodes. These putative dichotomies, beginning with nature–nurture and biology–culture (which, in turn, underpin BE–HCE and vocal–gestural and natural-language dichotomies), have been comprehensively rejected. The development of what we consider to be our uniquely human capacities is metaphysically realized in relationships that include continuously changing neural, corporeal, and sociocultural resources that are insensitive to putative boundaries between internal and external environments. "Biology" and "culture," it has been argued, interact continuously in social contexts to build such capacities. Relatedly, the view of emotion and language operating primarily in individual psychological economies has also been rejected; emotions and language both develop and operate primarily in social economies. In addition, and consistent with this idea, emotions should be construed as a unified natural class of essentially ostensive-expressive devices, the function of which is to enable human social life. Certainly, cognitions, feelings, and action tendencies are involved, and the well-being of the individual emoting subject is protected, but only because individual well-being contributes to, and is enabled by, normal human emotionality.[1]

In light of these more overarching results and rather more specifically, avenues for future research will usefully include further investigation into the nature of Scaffolding 1 and 2 and, relatedly, their place or potential place in education. They also include further investigation into the control of online interactivity in the world and the development of morality, the latter for reasons to be explained. I have touched on all these points in this monograph, and all deserve further detailed investigation. In addition, I suggest that they need investigation from an interdisciplinary or multidisciplinary perspective, with contributions from scholars in a range of relevant disciplines. This will enable the sort of synthesis I claimed that only interdisciplinary or multidisciplinary perspectives can achieve. This does not simply mean that scholars from one discipline trawl the relevant literatures to inform their theorizing, as, for example, I have done; it potentially provides opportunities for researchers from a range of disciplines who are interested in the same phenomena to collaborate from the beginning and on the hoof on research projects of mutual relevance. (This is not to deny that unidisciplinary research has its place but rather to suggest that phenomena as complex as human development and behavior in the world are probably not two of those places.)

8.2 Moral Development

In previous chapters, I have been at pains to argue that two highly complementary repertoires of constraints operating synchronously in a tight social and physical world provide the basis for the development of human emotionality and language (and, by implication, symbolic thought); that is, they provide the basis for what are commonly construed as uniquely human capacities. I argued that neonates' sensorimotor competencies, STBPs, and affect expressions, together with caregivers' intuitive parenting skills, constructed the very close, linguistically mediated social relationship that is prerequisite to full human development. I further argued that the development of both BEs and HCEs is language and culture dependent and that both are radically widely realized; that is, their metaphysical realization provides clear examples of DFI with CT. In addition, I argued that this linguistically dependent developmental trajectory resulted in a transformation from interpersonal emotional regulation to intrapersonal emotional regulation. The increasing use of language in requests, advice, explanations, and admonitions and praise (parents) and in action and feeling description and self-instruction (children) enables the transformation to intrapersonal emotional regulation. This transformation is accompanied by changes in locus of control in the total emotional realizer system.

Language development also depends on inbuilt neonatal and maternal constraints and, in particular, on motherese, the exaggeratedly succinct vocalizations that naturally accompany other activities and, when the child signals readiness, the introduction of deliberate language pedagogy. I characterized motherese as providing Scaffolding 2 to both emotional and language development, and deliberate pedagogy as providing Scaffolding 1. Language development is also radically widely realized and provides a further example of DFI with CT.

Human emotionality, language, and symbolic thought are uniquely human capacities, but they are not the only ones. Another uniquely human capacity, one that emotions, language, and thought inevitably feature in and ultimately enable, is that of moral agency. Given this, I think it plausible to suggest that children learn their ethics or morals through precisely the same mechanisms and at the same time as they learn to emote and to speak. Children learn to emote and to speak in culturally embedded interaction, and one dimension of such cultural embedding is moral or ethical. I think it plausible that just as Amy learns to say please and thank you for the reason that they are normal etiquette, so she learns not to squeeze Toby's ears for the reason that it will hurt Toby, and not to help herself to Sarah's pencil for the reason that it belongs to Sarah. If I am right, then moral development is radically widely realized, too, and moral agency is the result of a transformation from interpersonal to intrapersonal moral regulation with concomitant changes to locus of control in total realizer systems.

It would be interesting to investigate the extent to which emotional and language development influence moral development and the nature of the moral developmental trajectory. I think this is worth investigating using the same or similar method as I have used in this monograph. It is worth investigating for at least three different but related reasons. The first is that moral development, just like emotional and language development, is typically considered to be individualistic; both cognitivists and emotivists (both very generally construed) view moral development as part of individual psychological economies. If moral development can be shown to be developed concurrently with other radically widely realized uniquely human capacities, the case for DFI with CT in human development will be sealed. The second reason is that empirical evidence (Damasio 1994) suggests that they are related. The neocortex is involved in emotional regulation, and regulation takes time and a typical trajectory through which to develop. The third reason is that such a program would be timely. To begin with, the emotional–rational debate (yet another enduring putative dichotomy) has taken something of center stage again in mainstream moral psychology (Mackenzie 2012) with the positing of, inter alia, a language module (Hauser 2006) and social intuitionism (Haidt 2001), both of which deny rationality any fundamental role in moral agency. Thorough analysis of how children actually do learn to be moral agents may reconcile the perennial emotional–rational debate in moral theorizing in much the same way as my research on emotional ontogenesis has done with emotion theory.

8.3 Scaffolding

In addition, some conceptual ground clearance in terms of developmental and behavioral scaffolding would be useful. In 2003, Andy Clark suggested that the single most important task for a science of the biotechnological mind was to investigate the range and type of cognitive scaffolding and the different ways in which nonbiological scaffoldings can augment (or impair) performance on tasks.

The construction of the type of taxonomy Clark recommends is under way but requires refinement. R. Wilson and Clark (2009) distinguish technological, sociocultural, and natural environmental resources and suggest that they can be integrated either very temporally or almost permanently into cognitive structures. It would be useful to flesh out these categories with specific examples: what kinds of resources are functionally integrated into cognitive wholes, and what are their different effects on brains, behavior, affective states, and cognitive structure (Sutton 2010)? In addition, are they genuine examples of DFI with CT or examples of DFI or even only of "FI"? These questions relate to scaffolding.

Throughout the book, I have alluded to Scaffolding 1, that is, the assisted "walking" type of scaffolding or, as Sterelny (2010) would have it, beaver dams and termite mounds, and Scaffolding 2, the mutual structure-building type of scaffolding found in emotional and language development. These two types of scaffolding would appear to lie at opposite ends of a continuum, and if the idea of a continuum is correct, then a range of scaffoldings must exist, some of which will be clearly identifiable as one sort and some as the other. Scaffolding 2, I have claimed, results in the synchronous but unequal coconstruction of neural circuits in neonatal and maternal brains. This is structure building writ large and is relatively permanent (subject to some refinement, degeneration, and decay). Scaffolding 2 is coactive. Scaffolding 1, of the assisted "walking" and explicit pedagogy variety, involves an adult helping to build behavioral capacity in children, but inevitably involves some neuromuscular development. Scaffolding 1 of the beaver dam variety involves a physical construction built by generation *N* to facilitate the development and activities of generation *N* + 1. In both cases of Scaffolding 1, therefore, the support is unidirectional.

There will inevitably be scaffoldings, however, that will be difficult to classify, and this will be particularly true when development is sufficiently progressed to allow the complementing or replacing of Scaffolding 2 by Scaffolding 1. An example of this type of replacement is provided by language development, when motherese is complemented or replaced by explicit linguistic pedagogy. Some extremely subtle changes will occur, and apart from constructing the type of taxonomy Clark recommends, it would also be interesting to investigate what individual differences in both mothers and children and in physical and social environments influenced such changes.

8.4 Scaffolding 1 and Education

Given the wide availability of a range of exograms and systems for their use and manipulation, the role of deliberate scaffolding in schools and universities might usefully be explored. Examples of Scaffolding 1 involving computer-assisted learning could be incorporated into the full range of education communities to augment local face-to-face classroom education (if appropriately technologically resourced) and to promote computer literacy. Research into this potentially very fruitful education augmentation has already begun (e.g., Newstead et al. 2012), and it has begun to harvest some interesting and pedagogically salient results. For example, certain types of computer-assisted learning, such as the use of virtual worlds to enhance scientific understanding, result in more positive learning outcomes when introduced after individual student failure in the same tasks than when introduced before any individual task attempts (Newstead et al. 2012). It would be worth investigating whether, for example, prior failure motivates students to learn or somehow primes relevant neural circuits, or both. It is certainly interesting to note that caregiver-provided Scaffolding 1 in human development is generally provided because the infant or child cannot provide whatever she wants for herself; that is, she has either failed in her attempt or signaled to her caregiver her readiness for support. It has also been found that, in some cases of student collaboration on computer-assisted learning tasks, results are less positive (and easily measurable) learning outcomes. Clearly, therefore, the full range of potential scaffolds in differing educational contexts and their various strengths and limitations should usefully be explored.

8.5 Online Activity in the World

A further line of inquiry would be to explore the extent to which online activity in the world is radically widely realized. My research has mentioned utterance activity, mutual gazing, smiling, object passing, and visual perception as examples, but given the range of these, it is entirely plausible to suggest that there may well be others. There is also a distinction or perhaps even a relationship to be explored between interactivity that is scaffolded 2 and straightforward locational or vehicle externalism. Interactivities that are scaffolded 2 include utterance activity, mutual gazing, smiling, and object passing, where the activity of one interactor crucially depends on the activity of another. Visual perception is an example of straightforward locational externalism, where formal features in the environment are involved in wide computational processes. It has become obvious, however, during the development of this monograph that visual perception is very heavily involved in at least the examples of Scaffolding 2 given here. This makes sense. The function of our sensory-perceptual apparatuses is to keep track of events unfolding, moment to moment, in the real world so that we can respond appropriately. This implies that all our online interactivity is radically widely realized, and suggests that there might be auditory, olfactory, and tactile analogues to saccadic scanning in visual perception.[2] If this is the case, and it is certainly worth exploring, what remains for intracranialist processing contracts decisively. It seems that the only genuinely intracranial (but taxonomically externalist) processes are those when we are deliberately off-line, concentrating in terms of reflection, planning, reasoning, and so on. Yet even here, our subpersonal sensory-perceptual processes continue to operate to protect our interests.

8.6 Methodological Considerations

I suggested at the beginning of the chapter that the future research I recommend should be undertaken by interdisciplinary or multidisciplinary teams because this kind of approach allows important synthetic theorizing; it motivates new perspectives on relations between the cognitive and social sciences and the humanities (Sutton 2010). I think it fitting, given Andy Clark's pivotal and visionary leadership of the Extended Mind brigade, that he has almost the last word on methodological considerations:

> Much of what matters about human intelligence is hidden not in the brain, nor in the technology, but in the complex and iterated interactions and collaborations between the two. ... The study of these interaction spaces is not easy, and depends both on new multidisciplinary alliances and new forms of modelling and analysis. The pay-off, however, could be spectacular: nothing less than a new kind of cognitive scientific collaboration involving neuroscience, physiology, and social, cultural, and technological studies in equal measure. (A. Clark 2001, 154)

Clark's vision has become reality in some multidisciplinary collaborative research groups involving, inter alia, cognitive neuroscientists, cognitive psychologists, historians, linguists, and philosophers (e.g., Sutton 2012) working within an Extended Mind framework. Barnier (2012), for instance, a cognitive psychologist, has been involved in multidisciplinary research into memory for the last five years and claims that multidisciplinary research, undertaken from the ground up, requires the development of what she terms "transformative" methodologies. These methodologies, in turn, can harvest transformative insights in theory development. Barnier and colleagues do not claim that establishing such research groups or managing their ongoing operations is unproblematic. They insist, however, that when such research groups function as they should, the results they harvest can be unexpected and spectacular. For example, research has shown that 25 percent of partners, living at home in long-term relationships, who have a degree of neural decay normally diagnostic of Alzheimer's, remain symptom free. Barnier and colleagues have shown that this is because such subjects use their long-term partners as external memory sources. The potential therapeutic value of this research is clear. Research undertaken in other naturalistic settings, such as workplaces and schools, and informed by an Extended Mind framework has obvious potential to harvest other practically relevant results.

8.7 Concluding Summary

Human emotions develop from inborn emotional precursors, to basic emotions and thence to higher cognitive emotions; as such, they form a natural class in which precursor emotions are basic. Human emotions develop to enable human social life, initially through interpersonal control and subsequently through intrapersonal control. They develop through the continuous interaction of progressively fine-tuned neonatal or child affect expressions and caregiver intuitive parenting skills in a close, linguistically mediated social relationship. In addition, they develop concurrently through the same developmental mechanisms as language and, by implication, symbolic thought. The degree and complexity of functional integration of neural, corporeal, and sociocultural resources in human development are such that they represent a world-to-brain transcranialist achievement. Directions for future research include analyses of moral development. In addition, in light of the degree and complexity of functional integration of neural, corporeal, and sociocultural resources in human ontogenesis, research into human development and behavior in naturalistic settings should preferably be multidisciplinary. Finally, Ruth Millikan stated in one of her papers that she wanted to kill rationalism stone dead and then beat up on it. My aspirations for this monograph are similar. I hope that I have killed the biology–culture dichotomy in human development stone dead, and the intracranialist metaphysics that underpin it, and then beat up on them.

Notes

1 Introduction and Chapter Outlines

1. Research is under way to discover what expressions, gestures, etc., if any, are automatically imitated by human neonates. Early findings suggest that only tongue protrusion is automatically imitated, but imitation is significantly increased after training with an adult model: "TP (tongue protrusion) can be accounted for by associative learning and that early social experience improves newborn imitation. Whilst postnatal experience is not necessary for facial gesture imitation, neonatal imitation may nonetheless be influenced by experiences in the first weeks of life" (Slaughter, Oostenbroek, Suddendorf, and Nielson, personal communication, 2014).

2. I assume that coding is exclusively involved in the communication systems of many submammalian species, such as honeybees. The bee has the brain the size of a pinhead, which contains a mere 1,000 neurons.

3. I am not the only or the first theorist to suggest this idea. Menary (2007, 2009), for instance, in his Cognitive Integration Theory and, more especially, in his Enculturated Mind Theory (2013), does so too.

2 Theories of Emotion

1. Morton's views are so different from mainstream theories that I do not review them here. For those who enjoy good old-fashioned philosophy, however, they make very interesting reading.

2. A unified ontological class is defined by four characteristic functional roles: (i) a phenomenally encoded appraisal of environment or of the cognitive system itself; (ii) the preparation and motivation for action by eliciting physiological changes; (iii) a typical expression that signals the emotion and action readiness to other people; and (iv) a flexibility of behavior that is a function of degree of cognitive content involved. Emotions are not a natural kind, according to Zinck and Newen (2008), because they have different causal mechanisms.

3. Prinz (2004, 157) suggests "a highly speculative list" of BEs; these are frustration, panic, anxiety, physical disgust, separation distress, and aversive self-consciousness (all negative) and satisfaction, stimulation, and attachment (all positive).

4. Hohmann's findings have not gone unchallenged; see, e.g., Chwalisz, Diener, and Gallagher 1988.

5. Homologous traits differ from analogous traits in that analogues occur in different evolutionary lineages as a result of adapting to a common evolutionary problem, but share no common ancestor. The eyes of octopuses and the eyes of human beings are examples of analogous traits.

6. There are different construals of what emotion categories are basic emotions. Zinck and Newen (2008), for instance, exclude surprise and disgust, and Ekman (1994) adds contempt to the list.

7. These criteria were initially identified in the 1820s. A famous modern formulation is that of Remane (1952), reproduced in Reidl (1977) and translated by Jeffries. Clark, however, credits them to Ereshefski (2007), which provides a useful reference source.

3 Metaphysics and Mind

1. The latest, most radical, and most controversial form of externalism is extensive enactivism, due to Hutto, Kirchhoff, and Myin (2014), which I note briefly in chap. 2, sec. 2.7.3. It is controversial because it denies that cognition fundamentally involves contentful mental representations.

4 Mirror, Mirror ... Human Emotional Ontogenesis

1. But see my cautionary endnote 1 in chap. 1.

2. Not all smiling in neonates is a function of mimicry. Early research shows that blind and deaf and blind neonates smile, laugh (Eibl-Eibesfeldt 1973), and cry spontaneously, but their facial expressions differ from those of sighted infants. Blind infants show much more facial involvement or exaggerated facial activity than sighted children, and this activity increases with age (Fulcher 1942). Parents attempt to suppress such exaggerated and odd expressions (J. Thompson 1941), but despite their efforts, such facial expressions actually increase with age (Fulcher 1942). The reverse is the case for sighted children; facial activity decreases with age, but differentiation of emotion becomes sharper. The difference is thought to be due to sighted children's use of visual cues and mimicry (J. Thompson 1941; Fulcher 1942).

3. I am indebted to John Sutton for this insight.

5 Out of the Mouths of Babes and Sucklings

1. The timing of the onset of JVA remains controversial. Butterworth and coworkers construe it as emerging gradually from about four months, whereas others, such as Moore and Corkum (1994), construe its emergence as coincident with pointing comprehension at about one year. The reasons for the difference may be due, at least in part, to experimental conditions. For instance, evidence now suggests that early comprehension depends on an infant seeing the hand and target in proximal visual space and simultaneously (Butterworth 2013).

6 From Evolution to Emotionese

1. Millikan's is not the only teleological theory of function or one that is particularly idiosyncratic. Indeed, Papineau (1987) observes that in the early 1980s the time had come for such theories, and both he and Millikan recognized similarities in their positions. I favor Millikan because of her obvious sympathy for nonhuman (as well as human) species.

2. I use the word *design* metaphorically here and in what follows.

3. After Hugh Lofting's two-headed creature of the same name.

7 Loose Talk, Tight Worlds

1. Aristotle also subscribed to a substitution view; see sec. 7.2.2.

8 Once More, with Feeling

1. This is not to suggest that the human emotional developmental trajectory is exclusively positive. It is possible that during development, parental interests will conflict with those of offspring (e.g., Uller 2008), although a mutual sensitivity, much larger than evolutionary theory would predict, has also been posited (P. Bateson 1994). Much more research is required to identify the conditions under which parental effects will be positive and involve factors classified as "environmental" as well as parental (Uller 2008).

2. Some researchers are already focusing on other sensory modalities to investigate their enactive or embedded natures, with differing results. For example, Gray and Tanesani (2010) deny that taste is an enactive phenomenon, whereas I suggest that speech perception is an extended phenomenon.

References

Adams, F., and K. Aizawa. 2001. The bounds of cognition. *Philosophical Psychology* 14 (1): 43–64.

Adams, F., and K. Aizawa. 2005. Defending nonderived content. *Philosophical Psychology* 18: 661–669.

Adams, F., and K. Aizawa. 2008. *The Bounds of Cognition*. Oxford: Blackwell.

Adams, F., and K. Aizawa. 2009. Why the mind is still in the head. In *The Cambridge Handbook of Situated Cognition*, 9th ed., ed. P. Robbins and M. Aydede, 78–95. Cambridge: Cambridge University Press.

Adams, F., and K. Aizawa. 2010a. Defending the bounds of cognition. In *The Extended Mind*, ed. R. Menary, 67–80. Cambridge, MA: MIT Press.

Adams, F., and K. Aizawa. 2010b. The value of cognitivism in thinking about extended cognition. *Phenomenology and the Cognitive Sciences* 9 (4): 579–603.

Ainsworth, M. D. S., and S. M. Bell. 1974. Mother-infant interaction and the development of competence. In *The Growth of Competence*, ed. K. S. Connolly and J. S. Bruner, 97–118. New York: Academic Press.

Aktar, N., and M. Tomasello. 1998. Intersubjectivity in early language learning and use. In *Intersubjective Communication and Emotion in Early Ontogeny*, ed. S. Braten, 316–335. New York: Cambridge University Press.

Aristotle. 1984a. *Poetics*. In *The Complete Works of Aristotle*, rev. Oxford translation, ed. J. Barnes. Princeton, NJ: Princeton University Press.

Aristotle. 1984b. *Rhetoric*. In *The Complete Works of Aristotle*, rev. Oxford translation, ed. J. Barnes. Princeton, NJ: Princeton University Press.

Armon-Jones, C. 1986. The thesis of constructionism. In *The Social Construction of Emotion*, ed. R. Harre, 32–56. London: Oxford University Press.

Arnold, M. 1960. *The Nature of Emotion*. Harmondsworth: Penguin.

Averill, J. 1980. A constructionist view of emotion. In *Emotion: Theory, Research, and Experience*, vol. 1, ed. R. Plutchik and H. Kellerman, 305–336. New York: Academic Press.

Bach, K. 1982. De re belief and methodological solipsism. In *Thought and Object: Essays on Intentionality*, ed. A. Woodfield. Oxford: Oxford University Press.

Barnier, A. 2012. Otto and the barking dog: How the extended mind thesis helps us understand collaborative remembering. Paper presented at Distributed Cognition and Distributed Agency Workshop, March 16–18, Macquarie University, Sydney.

Barrett, K. C. 1995. A functionalist approach to shame and guilt. In *The Self-Conscious Emotions: The Psychology of Shame, Guilt, Embarrassment, and Pride*, ed. J. P. Tangney and K. W. Fischer, 25–63. New York: Guildford Press.

Barrett, L. F. 2005. Feeling is perceiving: Core affect and conceptualization in the experience of emotion. In *Emotions: Conscious and Unconscious*, ed. P. M. Niedenthal and P. Winkelman, 255–284. New York: Guilford Press.

Barrett, L. F. 2006. Solving the emotion paradox: Categorization and the experience of emotion. *Personality and Social Psychology Review* 10:20–46.

Barrett, L. F. 2009. Variety is the spice of life: A psychological construction approach to understanding variability in emotion. *Cognition and Emotion* 23:1284–1306.

Barrett, L. F. 2011. Was Darwin wrong about emotional expressions? *Advances in Psychological Sciences* 20 (6): 400–406.

Bateson, G. 1972. *Steps to an Ecology of Mind*. New York: Ballantine.

Bateson, M. C. 1975. Mother-infant exchanges: The epigenesis of conversational interaction. In *Developmental Psycholinguistics and Communication Disorders*, ed. D. Aaronson and R. W. Reiber, 107–113. New York: Academy of Sciences.

Bateson, M. C. 1979. The epigenesis of conversational interaction: A personal account of research development. In *Before Speech: The Beginning of Interpersonal Communication*, ed. M. Bullowa, 63–78. Cambridge: Cambridge University Press.

Bateson, P. 1994. The dynamics of parent-offspring relationships in mammals. *Trends in Ecology and Evolution* 10:399–403.

Beardsley, M. 1962. The metaphorical twist. *Philosophy and Phenomenological Research* 22 (3): 293–307.

Behne, T., M. Carpenter, and M. Tomasello. 2005. One-year-olds comprehend the communicative intentions behind gestures in a hiding game. *Developmental Science* 8 (6): 492–499.

Berti, A., and F. Frassinetti. 2000. When far becomes near: Remapping of space by tool use. *Journal of Cognitive Neuroscience* 12 (3): 415–420.

Black, M. 1954–55. Metaphor. *Proceedings of the Aristotelian Society* 55:273–294.

Black, M. 1993. More about metaphor. In *Metaphor and Thought*, ed. A. Ortony, 19–41. Cambridge: Cambridge University Press.

Blass, E. M., and V. Ciaramitaro. 1994. A new look at some old mechanisms in human newborns: Taste and tactile determinants of state, affect, and action. *Monographs of the Society for Research in Child Development* 59 (1): 1–8.

Bloom, L. 1991. *Language Development from Two to Three*. Cambridge: Cambridge University Press.

Bloom, L. 1993. *The Transition from Infancy to Language: Acquiring the Power of Expression*. New York: Cambridge University Press.

Bloom, L., and E. Tinker. 2001. *The Intentionality Model of Language Acquisition: Engagement, Effort, and the Essential Tension in Development*. Monographs of the Society for Research in Child Development, serial no. 267, vol. 66, no. 4. Oxford: Blackwell.

Boukydis, C. F. Z., and R. L. Burgess. 1982. Adult physiological response to infant cries: Effects of temperament of infant, parental status, and gender. *Child Development* 53 (5): 1291–1298.

Boyd, R. 1989. What realism implies and what it does not. *Dialectica* 43 (1–2): 5–29.

Boyd, R. 1993. Metaphor and theory: What is "metaphor" a metaphor for? In *Metaphor and Thought*, ed. A. Ortony, 481–532. Cambridge: Cambridge University Press.

Brandon, R. N. 1990. *Adaptation and Environment*. Princeton, NJ: Princeton University Press.

Bratman, M. E. 1992. Shared cooperative activity. *Philosophical Review* 101 (2): 327–341.

Brazelton, T. B. 1983. Precursors for the development of emotions in early infancy. In *Emotion: Theory, Research, and Experience*, vol. 2, ed. R. Plutchik and H. Kellerman, 35–55. San Diego, CA: Academic Press.

Brown, L. 1993. *New Shorter Oxford Dictionary*. Oxford: Clarendon Press.

Bruner, J. S. 1976. On prelinguistic prerequisites for speech. In *Recent Advances in the Psychology of Language: Language Development and Mother-Child Interaction*, ed. R. N. Campbell and R. T. Smith, 199–214. London: Plenum.

Buller, D. J. 1999a. Natural teleology. In *Function, Selection, and Design*, ed. D. J. Buller, 1–28. Albany, NY: SUNY Press.

Buller, D. J. 1999b. Etiological theories of function: A geographical survey. In *Function, Selection, and Design*, ed. D. J. Buller, 281–306. Albany, NY: SUNY Press.

Burge, T. 1979. Individualism and the mental. In *Midwest Studies in Philosophy*, vol. 4, ed. P. French and H. Wettstein, 73–121. Minneapolis: University of Minnesota Press.

Bush, D. E. A., G. E. Schafe, and J. Le Doux. 2009. The neural basis of fear functioning. In *Handbook of Neuroscience for the Behavioral Sciences*, vol. 2, ed. G. G. Bernston and J. J. Cacioppo, 762–780. Hoboken, NJ: John Wiley.

Bushnell, W. R. 2013. The origins of face perception. In *The Development of Sensory, Motor, and Cognitive Capacities in Early Infancy: From Sensation to Cognition*, ed. F. Simion and G. Butterworth, 69–86. Hove, England: Psychology Press.

Butler, D. L. 1998. In search of the architect of learning: A commentary on scaffolding as a metaphor for instructional interactions. *Journal of Learning Disabilities* 31:374–385.

Butterworth, G. 2008. Joint visual attention in infancy. In *Blackwell Handbook of Infant Development*, ed. G. Bremner and A. Fogel, 213–240. Oxford: Blackwell.

Butterworth, G. 2013. What is special about pointing in babies? In *The Development of Sensory, Motor, and Cognitive Capacities in Early Infancy: From Sensation to Cognition*, ed. F. Simion and G. Butterworth, 171–190. Hove, England: Psychology Press.

Butterworth, G., and I. Grover. 1989. Joint visual attention: Manual pointing and preverbal communication in human infancy. In *Attention and Performance XII*, ed. M. Jeannerod, 605–624. Hillsdale, NJ: Erlbaum.

Camras, L. A., R. Campos, J. Campos, K. Miyake, H. Oster, T. Ujiie, L. Wang, and Z. Meng. 1998. Production of emotional facial expressions in European American, Japanese, and Chinese infants. *Developmental Psychology* 34 (4): 616–628.

Cash, M. 2013. Cognition without borders: "Third Wave" society distributed cognition and relational autonomy. *Cognitive Systems Research* 25–26:61–71. doi:10.1016/j.cogsys.2013.03.007

Chappell, P. F., and L. W. Sander. 1979. Mutual regulation of neonatal-maternal interactive process: Context for the origins of communication. In *Before Speech: The Beginning of Interpersonal Communication*, ed. M. Bullowa. Cambridge: Cambridge University Press.

Chwalisz, K., F. Diener, and D. Gallagher. 1988. Autonomic arousal feedback and emotional experience: Evidence from the spinal cord injured. *Journal of Personality and Social Psychology* 54:820–828.

Clark, A. 1997. *Being There: Putting Brain, Body, and World Together Again*. Cambridge, MA: MIT Press.

Clark, A. 1998. Author's response: Review symposium on *Being There*. *Metascience* 7 (1): 95–103.

Clark, A. 2001. Reasons, robots, and the extended mind. *Mind and Language* 16: 121–145.

Clark, A. 2003. *Natural Born Cyborgs: Mind, Technologies, and the Future of Human Intelligence*. Oxford: Oxford University Press.

Clark, A. 2008. *Supersizing the Mind*. Oxford: Oxford University Press.

Clark, A. 2010. Coupling, constitution, and the cognitive kind: A reply to Adams and Aizawa. In *The Extended Mind*, ed. R. Menary, 81–100. Cambridge, MA: MIT Press.

Clark, A., and D. Chalmers. 1998. The extended mind. *Analysis* 58 (1): 7–19.

Clark, H. H. 1996. *Using Language*. Cambridge: Cambridge University Press.

Clark, J. 2010. Relations of homology between higher cognitive emotions and basic emotions. *Biology and Philosophy* 25 (1): 75–94.

Clark, R., R. Menna, and W. S. Manel. 2013. Maternal scaffolding and children's social skills: A comparison between aggressive preschoolers and non-aggressive preschoolers. *Early Child Development and Care* 183 (5): 707–725.

Cohen, T. 1976. Metaphor and the cultivation of intimacy. In *On Metaphor*, ed. S. Sachs, 1–11. Chicago: University of Chicago Press.

Colombetti, G. 2007. Enactive appraisal. *Phenomenology and the Cognitive Sciences* 6:527–546.

Colombetti, G. 2009a. From affect programs to dynamical discrete emotions. *Philosophical Psychology* 22 (4): 407–425.

Colombetti, G. 2009b. What language does to feelings. *Journal of Consciousness Studies* 16:4–26.

Colombetti, G. 2010. Enaction, sense-making, and emotion. In *Enaction: Toward a New Paradigm for Cognitive Science*, ed. J. Stewart, O. Gapenne, and E. A. Di Paolo, 145–164. Cambridge, MA: MIT Press.

Colombetti, G. 2011. Varieties of pre-reflective self-awareness: Foreground and background bodily feelings in emotion experience. *Inquiry* 54 (3): 293–313.

Colombetti, G. 2014. *The Feeling Body: Affective Science Meets the Enactive Mind*. Cambridge, MA: MIT Press.

Colombetti, G., and J. Krueger. 2015. Scaffoldings of the affective mind. *Philosophical Psychology*.

Colonnesi, C., G. J. J. M. Stams, I. Koster, and M. Noomb. 2010. The relation between pointing and language development: A meta-analysis. *Developmental Review* 30 (4): 352–366.

Cosmides, L., and J. Tooby. 2000. Evolutionary psychology and the emotions. In *Handbook of Emotions*, ed. M. L. Lewis and J. M. Haviland-Jones, 91–115. New York: Guildford Press.

Cott, H. B. 1966. *Adaptive Colouration in Animals*. London: Methuen.

Cowley, S. J. 1998. Of timing, turn-taking, and conversations. *Journal of Psycholinguistic Research* 27 (5): 541–571.

Cummins, R. 1975. Functional analysis. *Journal of Philosophy* 72 (20): 741–765.

Cummins, R. 1996. *Representations, Targets, and Attitudes*. Cambridge, MA: MIT Press.

Damasio, A. R. 1994. *Descartes' Error: Emotion, Reason, and the Human Brain*. New York: Grosset/Putnam.

Damasio, A. R., T. J. Grabowski, A. Bechara, H. Damasio, L. L. B. Ponto, J. Parvisi, and R. D. Hichwa. 2000. Subcortical and cortical brain activity during the feeling of self-generated emotions. *Nature Neuroscience* 3 (10): 1049–1056.

Damasio, A. R., D. Tranel, and H. Damasio. 1990. Individuals with sociopathic behaviour caused by frontal damage fail to respond automatically to social stimuli. *Behavioural Brain Research* 41 (2): 81–94.

Darwin, C. 1872/1965. *The Expression of Emotion in Man and Animals*. Chicago: University of Chicago Press.

Davidson, D. 1978. What metaphors mean. In *On Metaphor*, ed. S. Sachs, 29–46. Chicago: University of Chicago Press.

Dawkins, R. 1976. *The Selfish Gene*. Oxford: Oxford University Press.

Dawkins, R. 1986. *The Blind Watchmaker*. New York: W. W. Norton.

De Lancey, C. 2002. *Passionate Engines: What Emotions Reveal about Mind and Artificial Intelligence*. Oxford: Oxford University Press.

De Sousa, R. 1987. *The Rationality of Emotion*. Cambridge, MA: MIT Press.

De Sousa, Ronald. 2014. Emotion. In *The Stanford Encyclopedia of Philosophy* (spring 2014 ed.), ed. Edward N. Zalta. http://plato.stanford.edu.archives/spr2014/enties/emotion/.

Donald, M. 1991. *The Origins of the Modern Mind*. Cambridge: Cambridge University Press.

Dretske, F. 1981. *Knowledge and the Flow of Information*. Oxford: Blackwell.

Dretske, F. 1986. Misrepresentation. In *Belief: Form, Content, and Function*, ed. R. Bogdan, 17–36. Oxford: Oxford University Press.

Dretske, F. 1991. *Explaining Behavior: Reasons in a World of Causes*. Cambridge, MA: MIT Press.

Dunbar, R. I. M. 1996. *Grooming, Gossip, and the Evolution of Language*. London: Faber & Faber.

Eckerman, C. O. 1993. Imitation and toddlers' achievement of coordinated action with others. In *New Perspectives in Early Communicative Development*, ed. J. Nadel and L. Camioni, 116–138. London: Routledge.

Eibl-Eibesfeldt, I. 1973. Emotion behaviour of the deaf and blind born. In *Social Communication and Movement*, ed. M. Von Cranach and I. Vine, 163–194. New York: Academic Press.

Ekman, P. 1972. Universals and cultural differences in facial expression of emotions. In *Nebraska Symposium on Motivation*, ed. J. Cole, 34–84. London: Academic Press.

Ekman, P. 1984. Expression and the nature of emotion. In *Approaches to Emotion*, ed. K. R. Scherer and P. Ekman, 319–343. Hillsdale, NJ: Erlbaum.

Ekman, P. 1992. An argument for basic emotions. *Cognition and Emotion* 6 (3): 169–200.

Ekman, P. 1994. Moods, emotions, and traits. In *The Nature of Emotion: Fundamental Questions*, ed. P. Ekman and R. J. Davidson, 56–58. New York: Oxford: Oxford University Press.

Ekman, P. 2004. *Emotions Revealed: Understanding Faces and Feelings*. London: Phoenix.

Ekman, P., and W. Friesen. 1969. The repertoire of non-verbal behaviors: Categories, origins, usage, and coding. *Semiotica* 1:49–98.

Ereshefski, M. 2007. Psychological categories as homologies: Lessons from ethology. *Biology and Philosophy* 22 (5): 659–674.

Feldman, J. E. 2006. *From Molecule to Metaphor: A Neural Theory of Language*. Cambridge, MA: MIT Press.

Field, H. 1990. "Narrow" aspects of intentionality and the information-theoretic approach to content. In *Information, Semantics, and Epistemology*, ed. E. Villenueva, 102–116. New York: Blackwell.

Field, T. M., B. Healy, R. Greenburg, and D. Cohen. 1982. Discrimination and imitation of facial expressions by neonates. *Science* 218 (4568): 179–181.

Fischer, K. W., and T. Bidell. 1991. Constraining nativist inferences about cognitive capacities. In *The Epigenesis of Mind: Essays on Biology and Cognition*, ed. S. Carey and R. Gellman, 199–235. Hillsdale, NJ: Erlbaum.

Fodor, J. A. 1987. *Psychosemantics: The Problem of Meaning in the Philosophy of Mind.* Cambridge, MA: MIT Press.

Fogel, A. 1993. *Developing through Relationships.* Chicago: University of Chicago Press.

Fogel, A., B. J. King, and S. G. Shanker. 2008. Introduction: Why a dynamic systems approach to fostering human development. In *Human Development in the Twenty-first Century: Visionary Ideas from Systems Scientists,* ed. A. Fogel, B. J. King, and S. G. Shanker, 1–8. Cambridge: Cambridge University Press.

Frijda, N. H. 1986. *The Emotions.* New York: Cambridge University Press.

Fulcher, J. S. 1942. "Voluntary" facial expression in blind and seeing children. *Archives de Psychologie,* September.

Gergely, G., and J. Watson. 1999. Early social development: Contingency perception and the social biofeedback model. In *Early Social Cognition: Understanding Others in the First Months of Life,* ed. R. Rochat, 101–136. Mahwah, NJ: Erlbaum.

Gibbs, R. W., Jr. 2008. Metaphor and thought: The state of the art. In *The Cambridge Handbook of Metaphor and Thought,* ed. R. W. Gibbs Jr. Cambridge: Cambridge University Press.

Godfrey-Smith, P. 1991. Signal, detection, action. *Journal of Philosophy* 88 (12): 709–722.

Godfrey-Smith, P. 1994. A modern history theory of functions. *Noûs* 28 (3): 344–362.

Goldie, P. 2000. *The Emotions: A Philosophical Exploration.* Oxford: Oxford University Press.

Gordon, R. 1987. *The Structure of Emotions.* Cambridge: Cambridge University Press.

Gottlieb, G. 2002. *Individual Development and Evolution: The Genesis of Novel Behavior.* Mahwah, NJ: Erlbaum.

Gottlieb, G., and C. T. Halpern. 2008. Individual development as a system of coactions: Implications for research and policy. In *Human Development in the Twenty-first Century: Visionary Ideas from Systems Scientists,* ed. A. Fogel, B. J. King, and S. G. Shanker, 41–47. Cambridge: Cambridge University Press.

Granott, N. 1998. Unit of analysis in transit: From the individual's knowledge to the ensemble process. *Mind, Culture, and Activity* 5:42–66.

Gray, R., and A. Tanesani. 2010. Perception and action: The taste test. *Philosophical Quarterly* 60 (241): 718–734.

Greenspan, P. 2004. Practical reasoning and emotion. In *The Oxford Handbook of Rationality,* ed. A. Mele, 206–221. Oxford: Oxford University Press.

Greenwood, J. 2013. Is mind extended or scaffolded? Ruminations on Sterelny's (2010) extended stomach. *Phenomenology and the Cognitive Sciences* (Online First): 1–22. doi:10.1007/s11097-013-9337-8.

Greenwood, J., and A. Bonner. 2008. The role of theory-constitutive metaphor in nursing science. *Nursing Philosophy* 9 (3): 154–168.

Grice, H. P. 1957. Meaning. *Philosophical Review* 66 (3): 377–388.

Grice, H. P. 1989. Logic and conversation. In *Studies in the Way of Words*, ed. H. P. Grice, 1–143. Cambridge, MA: Harvard University Press.

Griffiths, P. E. 1992. Adaptive explanation and the concept of a vestige. In *Trees of Life: Essays in the Philosophy of Biology*, ed. P. E. Griffiths, 111–131. Dordrecht, the Netherlands: Kluwer Academic Publishers.

Griffiths, P. E. 1993. Functional analysis and proper function. *British Journal for the Philosophy of Science* 44:409–422.

Griffiths, P. E. 1997. *What Emotions Really Are*. London: University of Chicago Press.

Griffiths, P. E. 2003. Emotions. In *The Blackwell Guide to the Philosophy of Mind*, ed. S. Stich and T. Warfield, 288–308. Oxford: Blackwell.

Griffiths, P. E., and A. Scarantino. 2009. Emotion in the wild: The situated perspective on emotion. In *The Cambridge Handbook of Situated Cognition*, ed. P. Robbins and M. Aydede, 437–450. Cambridge: Cambridge University Press.

Grover, I. 1988. Comprehension of the pointing gesture in human infants. PhD diss., University of Southampton.

Gustafson, G. E., and K. L. Harris. 1990. Women's responses to young infants' cries. *Developmental Psychology* 26 (1): 144–152.

Guttenplan, S. 2004. *Objects of Metaphor*. Oxford: Oxford University Press.

Haidt, J. 2001. The emotional dog and its rational tail: A social intuitionist approach to moral judgment. *Psychological Review* 108 (4): 814–834.

Hanks, W. F. 2006. Joint commitment and common ground in a ritual event. In *Roots of Human Sociality: Culture, Cognition, and Interaction*, ed. N. J. Enfield and S. C. Levinson, 299–328. Oxford: Berg.

Harlow, H. F., and M. K. Harlow. 1962. Social deprivation in monkeys. *Scientific American* 207 (5): 136–146.

Harlow, H. F., and R. R. Zimmermann. 1959. Affectional responses in infant monkeys. *Science* 130 (3373): 421–432.

Harre, R. 1986. The social constructivist viewpoint. In *The Social Construction of Emotion*, ed. R. Harre, 2–14. London: Oxford University Press.

Harris, M. 2004. First words. In *Cognitive and Language Development in Children*, ed. J. Oates and A. Grayson, 63–106. Milton Keynes: Open University Press.

Hauser, M. C. 2006. *Moral Minds: How Nature Designed Our Universal Sense of Right and Wrong*. New York: Ecco.

Hebb, D. O. 1946. *Textbook of Psychology*. Philadelphia: W. B. Saunders.

Hendriks-Jansen, H. 1996. *Catching Ourselves in the Act: Situated Activity, Interactive Emergence, Evolution, and Human Thought*. Cambridge, MA: MIT Press.

Ho, D. 1986. Chinese patterns of socialisation: A review. In *The Psychology of Chinese People*, ed. M. Bond, 1–37. Oxford: Oxford University Press.

Hofstede, G. H. 2000. *Culture's Consequences: Comparing Values, Behaviors, Institutions, and Organizations across Nations*, 2nd ed. Thousand Oaks, CA: Sage.

Hohmann, G. W. 1966. Some effects of spinal cord lesions on experienced emotion feelings. *Psychophysiology* 3:143–156.

Holodynski, M., and W. Friedlmeier. 2006. *Development of Emotions and Emotion Regulation*. New York: Springer.

Holodynski, M., and R. Oerter. 2002. Motivation, emotion und handlungssteuerung. In *Entwicklungspsychologie*, ed. R. Oerter and L. Montada, 551–589. Weinheim, Germany: Psychologie Verlags Union.

Hull, D. L. 1974. *Philosophy of Biological Science*. Englewood Cliffs, NJ: Prentice Hall.

Hurley, S. 1998. *Consciousness in Action*. Cambridge, MA: Harvard University Press.

Husserl, E. 1952/1989. *Ideas Pertaining to a Pure Phenomenology and to a Phenomenological Philosophy: Second Book*. Trans. and ed. R. Rojcewicz and A. Schuwer. Dordrecht: Kluwer.

Hutto, D. D., and E. Myin. 2013. *Radicalizing Enactivism*. Cambridge, MA: MIT Press.

Hutto, D. D., M. D. Kirchhoff, and E. Myin. 2014. Extensive enactivism: Why keep it all in? *Frontiers in Human Neuroscience* 8 (706): 1–11.

Izard, C. E. 1978. On the ontogenesis of emotions and emotion-cognition relationships in infancy. In *The Development of Affect*, ed. M. Lewis and L. Rosenblum, 389–413. New York: Plenum.

Izard, C. E., and B. P. Ackerman. 2000. Motivational, organizational, and regulatory functions of discrete emotions. In *Handbook of Emotions*, ed. M. L. Lewis and J. M. Haviland-Jones, 253–264. New York: Guildford Press.

James, W. 1884. What is an emotion? *Mind* 9:188–205.

Johnson, D. M. 2003. *How History Made the Mind*. Chicago: Open Court.

Johnson, M. 1981. Introduction: Metaphor in philosophical tradition. In *Philosophical Perspectives on Metaphor*, ed. M. Johnson, 3–47. Minneapolis: University of Minneapolis Press.

Johnston, T. J. 2008. Genes, experience, behaviour. In *Human Development in the Twenty-first Century*, ed. A. Fogel, B. J. King, and S. G. Shanker, 18–24. Cambridge: Cambridge University Press.

Johnston, T. J., and L. Edwards. 2002. Genes, interaction, and the development of behaviour. *Psychological Review* 10 (9): 36–54.

Kaye, K. 1982. Organism, apprentice, and person. In *Social Interchange in Infancy*, ed. E. Z. Tronick, 183–196. Baltimore: University Park Press.

Keijzer, F., and M. Schouten. 2007. Embedded cognition and mental causation: Setting empirical boundaries on metaphysics. *Synthese* 158 (1): 109–125.

Kenny, A. 1963. *Action, Emotion, and Will*. London: Routledge & Kegan Paul.

Kirchhoff, M. D. 2012. Extended cognition and fixed properties: Steps to a third wave version of extended cognition. *Phenomenology and the Cognitive Sciences* 11 (2): 287–308.

Kirsch, P., C. Esslinger, Q. Chen, D. Mier, S. Lis, S. Siddhanti, H. Gruppe, B. Gallhofer, and A. Meyer-Lindenberg. 2005. Oxytocin modulates neural circuitry for social cognition and fear in humans. *Journal of Neuroscience* 25 (49): 11489–11493.

Kisilevsky, B. S., S. M. Hains, C. A. Brown, C. T. Lee, B. Cowperthwaite, S. S. Stutzmann, and M. L. Swansburg. 2009. Fetal sensitivity to properties of maternal speech and language. *Infant Behavior and Development* 32 (1): 59–71.

Kleinman, A. 1980. *Patients and Healers in the Context of Culture*. Berkeley: University of California Press.

Knowles, M., and R. Moon. 2006. *Introducing Metaphor*. London: Routledge.

Kovecses, Z. 1990. *Emotion Concepts*. New York: Springer.

Kripke, S. A. 1980. *Naming and Necessity*. Cambridge, MA: Harvard University Press.

Krueger, J. 2013. Ontogenesis of the socially extended mind. *Cognitive Systems Research*. doi:10.1016/j.cpgsys.2013.03.001.

Lakoff, G. 1993. The contemporary theory of metaphor. In *Metaphor and Thought*, ed. A. Ortony, 202–256. Cambridge: Cambridge University Press.

Lakoff, J., and M. Johnson. 1980. Conceptual metaphor in everyday language. *Journal of Philosophy* 77 (8): 453–486.

Lange, C. G. 1885/1922. The emotions: A psychophysical study. In *The Emotions by Carl Georg Lange and William James*, ed. K. Dunlop. Baltimore: Williams & Wilkins.

Langsdorf, P., C. E. Izard, M. Rayias, and E. A. Hembree. 1983. Interest expression, visual fixation, and heart rate changes in 2- to 8-month-old infants. *Developmental Psychology* 19 (3): 375–386.

Lazarus, R. S. 1991. *Emotion and Adaptation.* New York: Oxford University Press.

Le Doux, J. E. 1998. *The Emotional Brain.* New York: Simon & Schuster.

Le Doux, J. E. 2000. Emotion circuits in the brain. *Annual Review of Neuroscience* 23 (1): 155–184.

Ledwig, M. 2006. *Emotions: Their Rationality and Consistency.* New York: Peter Lang.

Legerstee, M., and J. Varghese. 2001. The role of maternal affect mirroring on social expectancies in three-month-old infants. *Child Development* 72 (5): 1301–1313.

Lester, B. M. 1984. Infant crying and the development of communication. In *The Psychobiology of Affective Development,* ed. N. Fox and R. J. Davidson, 231–258. Hillsdale, NJ: Erlbaum.

Levinson, S. C. 2006. On the human "interaction engine." In *Roots of Human Sociality: Culture, Cognition, and Interaction,* ed. N. J. Enfield and S. C. Levinson, 39–69. Oxford: Berg.

Lewis, D. 1975/1996. Languages and language. In *Readings in Language and Mind,* ed. H. Geirsson and M. Losonsky, 134–155. Oxford: Blackwell.

Lewis, M. D. 2005. Bridging emotion theory and neurobiology through dynamic systems modeling. *Behavioral and Brain Sciences* 28:169–245.

Lewis, M. D., and I. Granic. 2000. Introduction to *Emotion, Development, and Self-Organization: Dynamic Systems Approaches to Emotional Development,* by M. D. Lewis and I. Granic, 1–14. Cambridge: Cambridge University Press.

Lewis, M. M. 2013. *Infant Speech: A Study of the Beginning of Language.* Hoboken, NJ: Taylor & Francis.

Lickliter, R. 2008. Developmental dynamics: The view from the life sciences. In *Human Development in the Twenty-first Century: Visionary Ideas from Systems Scientists,* ed. A. Fogel, B. J. King, and S. G. Shanker. Cambridge: Cambridge University Press.

Lindquist, K., T. D. Wager, H. Kober, E. Bliss-Moreau, and L. F. Barrett. 2012. The brain basis of emotion: A meta-analytic view. *Behavioral and Brain Sciences* 35: 121–202.

Liskowski, U., M. Carpenter, A. Henning, T. Striano, and M. Tomasello. 2004. Twelve month olds point to share attention and interest. *Developmental Science* 73:297–307.

Liskowski, U., M. Carpenter, T. Striano, and M. Tomasello. 2006. 12- and 18-month-olds point to provide information for others. *Journal of Cognition and Development* 72:173–187.

Loar, B. 1995. Meaning. In *The Cambridge Dictionary of Philosophy*, ed. R. Audi, 471–476. Cambridge: Cambridge University Press.

Lowe, J., S. J. Erickson, P. MacLean, S. W. Duvall, R. K. Ohls, and A. F. Duncan. 2014. Associations between maternal scaffolding and executive functioning in 3- and 4-year-olds born very low birth weight and normal birth weight. *Early Human Development* 90:587–593.

Lowe, J. R., S. J. Erickson, P. MacLean, R. Schrader, and J. Fuller. 2013. Association of maternal scaffolding to maternal education and cognition in toddlers born preterm and full term. *Acta Paediatrica* 102 (1): 72–77.

Luria, A. R. 1961. Experimentelle analyse der entwicklung willensmassiger handlungen bei kindern. In *Perspectiven der Personlichkeitsforschung*, ed. J. C. Brengelman and H. P. David, 107–115. Bern: Huber.

Luria, A. R. 1980. *Higher Cortical Functions in Man*. New York: Basic Books.

Lutz, C. 1986. The domain of emotion words on Ifaluk. In *The Social Construction of Emotions*, ed. R. Harre, 267–288. Oxford: Blackwell.

Lyons, W. 1980. *Emotion*. Cambridge: Cambridge University Press.

Maccoby, E. E. 1992. The role of parents in the socialisation of children: An historical overview. *Developmental Psychology* 28 (6): 1006–1017.

Mackenzie, C. 2012. Emotions, reflection, and moral agency. In *Emotions, Imagination, and Moral Reasoning*, ed. R. Langdon and C. Mackenzie, 237–256. Hove, England: Psychology Press.

Maclean, P. D. 1990. *The Triune Brain in Evolution: Role in Paleocerebral Functions*. New York: Plenum.

Malatesta, C. Z., and C. E. Izard. 1984. The ontogenesis of human social signals: From biological imperative to symbol utilization. In *The Psychology of Affective Development*, ed. R. A. Fox and D. J. Davidson, 161–206. Hillsdale, NJ: Erlbaum.

Malatesta, C. Z., and A. Wilson. 1988. Emotion, cognition, interaction in personality development: A discrete emotions fundamentalist analysis. *British Journal of Social Psychology* 27:91–112.

Maravita, A., and A. Iriki. 2004. Tools for the body schema. *Trends in Cognitive Sciences* 8 (2): 79–86.

Mascolo, M. F. 2005. Change processes in development: The concept of coactive scaffolding. *New Ideas in Psychology* 23 (3): 185–196.

Mascolo, M. F., and K. W. Fischer. 2005. Constructivist theories. In *The Cambridge Encyclopedia of Child Development*, ed. B. Hopkins, R. G. Barr, G. Michel, and P. Rochat, 40–63. Cambridge: Cambridge University Press.

Mascolo, M. F., and S. Griffin. 1998. Alternative trajectories in the development of anger-related appraisals. In *What Develops in Emotional Development?* ed. M. F. Mascolo and S. Griffin. New York: Plenum.

McDonough, L. 2002. Basic level nouns: First learned but misunderstood. *Journal of Child Language* 29 (2): 357–377.

Meltzoff, A. N., and M. K. Moore. 1988. The origins of imitation in infancy: Paradigm, phenomena and theories. In *Advances in Infancy Research,* ed. C. Rovec-Collier and L. P. Lipsitt, 265–301. Norwood, NJ: Ablex.

Meltzoff, A. N., and M. K. Moore. 1989. Imitation in newborn infants: Explaining the range of gestures imitated and underlying mechanism. *Developmental Psychology* 25 (6): 954–962.

Menary, R. 2007. *Cognitive Integration: Mind and Cognition Unbounded.* Basingstoke: Palgrave Macmillan.

Menary, R. 2009. Intentionality, cognitive integration and the continuity thesis. *Topoi* 28:31–43.

Menary, R. 2010. Cognitive integration and the extended mind. In *The Extended Mind,* ed. R. Menary, 227–244. Cambridge, MA: MIT Press.

Menary, R. 2013. Cognitive integration, enculturated cognition, and the socially extended mind. *Cognitive Systems Research* 25–26:26–34. doi:10.1016/j.cogsys.2013.05.002.

Merleau-Ponty, M. 1962. *The Phenomenology of Perception.* Trans. and ed. C. Smith. New York: Humanities Press.

Miller, J. L., and P. W. Jusczyk. 1989. Seeking the neurobiological bases of speech perception. *Cognition* 33 (1): 111–137.

Millikan, R. G. 1984. *Language, Thought, and Other Biological Categories.* Cambridge, MA: MIT Press.

Millikan, R. G. 1989a. In defence of proper functions. *Philosophy of Science* 56 (2): 288–302.

Millikan, R. G. 1989b. Biosemantics. *Journal of Philosophy* 86 (6): 281–297.

Millikan, R. G. 1993. *White Queen Psychology and Other Essays for Alice.* Cambridge, MA: MIT Press.

Millikan, R. G. 1996. Pushmi-pullyu representations. *Philosophical Perspectives* 9: 185–200.

Millikan, R. G. 1998. Language conventions made simple. *Journal of Philosophy* 95 (4): 161–180.

Millikan, R. G. 2000. *On Clear and Confused Ideas*. Cambridge: Cambridge University Press.

Millikan, R. G. 2002. Biofunctions: Two paradigms. In *Functions: New Readings in the Philosophy of Psychology and Biology*, ed. R. Cummins, A. Ariew, and M. Perlman, 113–143. Oxford: Oxford University Press.

Millikan, R. G. 2004. *Varieties of Meaning*. The Jean Nicod Lectures, 2002. Cambridge, MA: MIT Press.

Millikan, R. G. 2005. *Language: A Biological Model*. Oxford: Oxford University Press.

Millikan, R. G. 2009. Biosemantics. In *The Oxford Handbook of Philosophy of Mind*, ed. B. P. McLaughlin, A. Beckerman, and S. Walter, 394–406. Oxford: Clarendon Press.

Moore, C., and V. Corkum. 1994. Social understanding at the end of the first year of life. *Developmental Review* 14 (4): 349–372.

Morelock, M. J., P. M. Brown, and A. M. Morrissey. 2003. Pretend play and maternal scaffolding: Comparisons of toddlers with advanced development, typical development and hearing impairment. *Roeper Review* 26 (1): 41–51. doi:10.1080/02783190309554238.

Morsbach, H., and W. J. Tyler. 1986. A Japanese emotion: Amae. In *The Social Construction of Emotions*, ed. R. Harre, 289–307. Oxford: Blackwell.

Morton, A. 2013. *Emotion and Imagination*. Cambridge: Polity.

Nash, R. A. 1989. Cognitive theories of emotion. *Noûs* 23 (4): 481–504.

Neisser, U. 1981. *Memory Observed: Remembering in Natural Contexts*. San Francisco: W. H. Freeman.

Nelson, E. E., and J. Panksepp. 1998. Brain substrates of infant-mother attachment: Contributions of opioids, oxytocin, and norepinephrine. *Neuroscience and Biobehavioral Reviews* 22 (3): 437–452.

Nelson, K., ed. 1989. *Narratives from the Crib*. Cambridge, MA: Harvard University Press.

Newstead, A., M. Jacobson, D. Richards, and C. A. Taylor. 2012. Collaborative virtual world for enhancing collective scientific understanding. Paper presented at Distributed Cognition and Distributed Agency Workshop, March 16–18, Macquarie University, Sydney.

Noveck, I. A., and D. Sperber. 2004. Introduction. In *Experimental Pragmatics*, ed. I. A. Noveck and D. Sperber, 1–24. Basingstoke: Palgrave Macmillan.

Nussbaum, M. C. 2001. *Upheavals of Thought: The Intelligence of the Emotions*. Cambridge: Cambridge University Press.

Oatley, K. 2004. *Emotions: A Brief History*. Oxford: Blackwell.

Oatley, K., and K. Johnson-Laird. 1987. Towards a cognitive theory of emotion. *Cognition and Emotion* 1 (1): 29–50.

Ohman, A. 1986. Face the beast and fear the face: Animal and social fears as prototypes for evolutionary analyses of emotions. *Psychophysiology* 23 (2): 123–145.

O'Regan, J. K., and A. Noe. 2001. A sensori-motor account of vision and visual consciousness. *Behavioral and Brain Sciences* 24 (5): 939–973.

Origgi, G., and D. Sperber. 2000. Evolution, communication, and the proper function of language. In *Evolution and the Human Mind: Modularity, Language, and Metacognition*, ed. P. Carruthers and A. Chamberlain, 140–169. Cambridge: Cambridge University Press.

Ortony, A., G. L. Clore, and A. Collins. 1988. *The Cognitive Structure of Emotions*. Cambridge: Cambridge University Press.

Ortony, A., and W. Turner. 1990. What's basic about basic emotions? *Psychological Review* 97 (3): 315–331.

Panksepp, J. 1998. *Affective Neuroscience*. New York: Oxford University Press.

Panksepp, J. 2000. Emotions as natural kinds within the mammalian brain. In *Handbook of Emotions*, 2nd ed., ed. M. Lewis and J. Haviland-Jones, 137–156. New York: Guildford Press.

Panksepp, J. 2007. The neuroevolutionary and neuroaffective psychobiology of the prosocial brain. In *The Oxford Handbook of Evolutionary Psychology*, ed. R. I. M. Dunbar and L. Barrett, 145–162. Oxford: Oxford University Press.

Papanicolaou, A. C. 1989. *Emotion: A Reconsideration of the Somatic Theory*. New York: Gordon and Breach Science Publishers.

Papineau, D. 1987. *Reality and Representation*. Oxford: Blackwell.

Papousek, H., and M. Papousek. 1987. Intuitive parenting: A dialectic counterpart to the infant's integrative competence. In *Handbook of Infant Development*, ed. J. D. Osofsky, 669–720. New York: John Wiley.

Papousek, H., and M. Papousek. 1999. Symbolbildung, emotionsregulation und soziale interaktion. In *Emotionale entwicklung, funktion, regulation und sociokultureller kontext von emotionen*, ed. W. Friedlmeier and M. Holodynski, 135–155. Heidelberg: Spektrum.

Papousek, M. 1990. Affektive verhaltensregulation des sauglings in Der eltern- kind-interaktion, ed. M. J. S. Pachler and H. M. Strabburg, 203–221. Lubeck, Germany: Hansisches Verlagskontor.

Papousek, M. 1994. *Vom ersten schrei zum ersten wor: Anfange der sprachentwicklung in der vorsspechlichen kommunikation*. Bern: Huber.

Pea, R. D. 2004. The social and technological dimensions of scaffolding and related theoretical concepts for learning, education and human activity. *Journal of the Learning Sciences* 13 (3): 423–452.

Perovic, S., and L. Radenovic. 2011. Fine-tuning nativism: The "nurtured nature" and innate cognitive structures. *Phenomenology and the Cognitive Sciences* 10 (3): 399–417.

Pinker, S. 1994. *The Language Instinct*. New York: HarperCollins.

Plato. 1974. *The Republic*. Trans. D. Lee. London: Penguin Books.

Polger, T. W. 2004. *Natural Minds*. Cambridge, MA: MIT Press.

Polger, T. W. 2007. Realization and the metaphysics of mind. *Australasian Journal of Philosophy* 85 (2): 233–259.

Povinelli, D. 2004. Behind the ape's appearance: Escaping anthropocentrisms in the study of other minds. *Daedalus* 133 (1): 29–41.

Povinelli, D. J., and T. J. Eddy. 1996. Factors influencing young chimpanzees' (Pan troglodytes) recognition of attention. *Journal of Comparative Psychology* 110 (4): 336–345.

Prechl, H. F. R. 1993. Principles of early motor development in the human. In *Motor Development in Early and Later Childhood*, ed. A. Kalverboer, B. Hopkins, and R. Geuze, 35–50. Cambridge: Cambridge University Press.

Price, C. 2001. *Functions in Mind: A Theory of Intentional Content*. Oxford: Oxford University Press.

Prinz, J. 2004. *Gut Reactions: A Perceptual Theory of Emotion*. Oxford: Oxford University Press.

Punter, D. 2007. *Metaphor*. London: Routledge.

Putnam, H. 1975. The meaning of "meaning." In *Language, Mind, and Knowledge*, ed. K. Gunderson, 131–193. Minneapolis: University of Minnesota Press.

Reddy, M. J. 1979/1993. The conduit metaphor: A case of frame conflict in our language about language. In *Metaphor and Thought*, ed. A. Ortony, 164–201. Cambridge: Cambridge University Press.

Reidl, R. 1977. *Order in Living Organisms: A Systems Analysis of Evolution*. Trans. and ed. R. P. S. Jeffries. Chichester: John Wiley.

Remane, A. 1952. *Die grundlagen des naturlichen systems, der vergleichenden anatomis und der phylogenetik*. Konigstein: Otto Koeltz.

Rensink, R. A., J. K. O'Regan, and J. J. Clark. 2000. On the failure to detect changes in scenes across brief interruptions. *Visual Cognition* 7 (1–3): 127–146.

Richards, I. A. 1936. *The Philosophy of Rhetoric*. New York: Oxford University Press.

Rime, B., S. Corsini, and G. Herbette. 2002. Emotion, verbal expression and the social sharing of emotion. In *The Verbal Communication of Emotions: Interdisciplinary Perspectives*, ed. S. R. Fussell, 185–208. Mahwah, NJ: Erlbaum.

Robbins, P., and M. Aydede. 2009. A short primer on situated cognition. In *The Cambridge Handbook of Situated Cognition*, ed. P. Robbins and M. Aydede, 3–10. Cambridge: Cambridge University Press.

Robinson, J. B., B. M. Burns, and D. Winders Davis. 2009. Maternal scaffolding and attention regulation in children living in poverty. *Journal of Applied Developmental Psychology* 30:82–91.

Rogoff, B. 1990. *Apprenticeship in Thinking: Cognitive Development in Social Context*. New York: Oxford University Press.

Rogoff, B., and J. Wertsch, eds. 1984. *Children's Learning in the "Zone of Proximal Development."* New Directions for Child Development, No. 23. San Francisco: Jossey-Bass.

Rothbaum, F., and G. Morelli. 2005. Attachment and culture: Directions for future research. In *Human Development and Culture: The Importance of Cross-Cultural Research to Social Sciences*, ed. W. Friedlmeier, P. Chakkarath, and B. Schwartz, 1–124. Lisse: Swets & Zeitlinger.

Rowlands, C. 2014. *Understanding Language Acquisition*. Abingdon, Oxford: Routledge.

Rowlands, M. 2003. *Externalism: Putting Mind and World Back Together Again*. London: Acumen.

Rowlands, M. 2006. *Body Language: Representation in Action*. Cambridge, MA: MIT Press.

Rupert, R. 2004. Challenges to the hypothesis of extended cognition. *Journal of Philosophy* 101 (8): 389–428.

Rupert, R. 2010. Representation in extended cognitive systems. In *The Extended Mind*, ed. R. Menary, 325–354. Cambridge, MA: MIT Press.

Scherer, K. R. 1994. Emotion serves to decouple stimulus and response. In *The Nature of Emotion*, ed. P. Ekman and K. T. Davidson, 127–130. New York: Oxford University Press.

Schore, A. N. 1997. Early organization of the non-linear right brain and development of a predisposition to psychiatric disorders. *Development and Psychopathology* 9:595–631.

Schwartz, P. H. 2002. The continuing usefulness account of proper function. In *Functions: New Essays in the Philosophy of Psychology and Biology*, ed. A. Ariew, R. Cummins, and M. Perlman, 244–260. Oxford: Oxford University Press.

Searle, J. 1979. Metaphor. In *Metaphor and Thought*, ed. A. Ortony. New York: Cambridge University Press.

Searle, J. 1983. *Intentionality: An Essay in the Philosophy of Mind*. New York: Cambridge University Press.

Shakespeare, William. 1997. *The Complete Works of William Shakespeare: The Alexander Text*. Ed. Peter Alexander. London: Collins.

Shanker, S. G. 2008. The impact of emotions and the emotional impact of a child's first words. In *Human Development in the Twenty-first Century: Visionary Ideas from Systems Scientists*, ed. A. Fogel, B. J. King, and S. G. Shanker, 65–71. New York: Cambridge University Press.

Shapiro, L. 2004. *The Mind Incarnate*. Cambridge, MA: MIT Press.

Shoemaker, S. 1984. Some varieties of functionalism. In *Identity, Cause, and Mind*, ed. S. Shoemaker, 261–286. New York: Cambridge University Press.

Simion, F., L. Regolin, and H. Bulf. 2008. A predisposition for biological motion in the newborn baby. *Proceedings of the National Academy of Sciences of the United States of America* 105 (2): 809–813.

Simons, D. J., and D. T. Levin. 1997. Change blindness. *Trends in Cognitive Sciences* 1 (7): 261–267.

Simons, D. J., and D. T. Levin. 1998. Failure to detect changes to people during a real world interaction. *Psychonomic Bulletin and Review* 5 (4): 644–649.

Solomon, R. C. 1977. *The Passions*. New York: Anchor Books.

Solomon, R. C. 1993. *The Passions: Emotions and the Meaning of Life*. Indianapolis: Hackett.

Solomon, R. C. 2003. *Not Passion's Slave*. Oxford: Oxford University Press.

Sperber, D., and D. Wilson. 1986. Loose talk. *Aristotelian Society Proceedings* 86: 153–172.

Sperber, D., and D. Wilson. 1995. *Relevance: Communication and Cognition*. 2nd ed. Oxford: Blackwell.

Sperber, D., and D. Wilson. 1998. The mapping between the mental and the public lexicon. In *Thought and Language*, ed. P. Carruthers and J. Boucher, 184–200. Cambridge: Cambridge University Press.

Sperber, D., and D. Wilson. 2005. Pragmatics. In *The Oxford Handbook of Contemporary Philosophy*, ed. F. Jackson and M. Smith, 468–504. Oxford: Oxford University Press.

Sperber, D., and D. Wilson. 2008. A deflationary theory of metaphors. In *The Cambridge Handbook of Metaphor and Thought*, ed. R. W. Gibbs Jr., 84–108. Cambridge: Cambridge University Press.

Spivey, M., and D. Richardson. 2009. Language processing embodied and embedded. In *The Cambridge Handbook of Situated Cognition*, ed. P. Robbins and M. Aydede, 382–400. New York: Cambridge University Press.

Spurrett, D., and S. Cowley. 2010. The extended infant: Utterance-activity and distributed cognition. In *The Extended Mind*, ed. R. Menary, 295–324. Cambridge, MA: MIT Press.

Sroufe, L. A. 1996. *Emotional Development: The Organization of Emotional Life in the Early Years*. New York: Cambridge University Press.

Stanford, W. B. 1936. *Greek Metaphor: Studies in Theory and Practice*. Oxford: Blackwell.

Sterelny, K. 2010. Minds: Extended or scaffolded? *Phenomenology and the Cognitive Sciences* 9 (4): 465–481.

Stern, D. N. 1985. *The Interpersonal World of the Infant: A View from Psychoanalysis and Development*. New York: Basic Books.

Sternberg, C. R., and J. J. Campos. 1990. The development of anger expressions in infancy. In *Psychological and Biological Approaches to Emotion*, ed. N. L. Stein, B. Leventhal, and T. Trabasso, 247–282. Hillsdale, NJ: Erlbaum.

Sutton, J. 2010. Exograms and interdisciplinarity: History, the extended mind, and the civilizing process. In *The Extended Mind*, ed. R. Menary, 189–226. Cambridge, MA: MIT Press.

Sutton, J. 2012. Nudge or coup? Different ways to radicalise distributed cognition. Paper presented to Distributed Cognition and Distributed Agency Workshop, March 16–18, Macquarie University, Sydney.

Sutton, J., C. B. Harris, P. G. Keil, and A. J. Barnier. 2010. The psychology of memory, extended cognition, and socially distributed remembering. *Phenomenology and the Cognitive Sciences* 9:521–560.

Thelen, E. 1981. Rhythmical behaviour in infancy: An ethological perspective. *Developmental Psychology* 18 (5): 760–775.

Thelen, E. 1984. Learning to walk: Ecological demands and phylogenetic constraints. In *Advances in Infancy Research,* ed. L. P. Lippsitt and C. Rovier-Collier, 213–260. Norwood: Ablex.

Thompson, J. 1941. Development of facial expression of emotion in blind and seeing children. *Archives of Psychology* 264. New York, July 1941.

Thompson, R. A. 1990. Emotion and self-regulation. In *Nebraska Symposium on Motivation,* ed. R. A. Thompson, 367–467. Lincoln: University of Nebraska Press.

Tomasello, M. 1999. *The Cultural Origins of Human Cognition.* Cambridge, MA: Harvard University Press.

Tomasello, M., J. Call, and A. Gluckman. 1997. A comprehension of novel communication signs by apes and human children. *Child Development* 68 (6): 1067–1080.

Tomasello, M., and M. Carpenter. 2005. The emergence of social cognition in three young chimpanzees. *Monographs of the Society for Research in Child Development* 70 (1): 1–152.

Trevarthen, C. 1979. Communication and cooperation in early infancy: A description of primary intersubjectivity. In *Before Speech: The Beginning of Interpersonal Communication,* ed. M. Bullova, 321–347. Cambridge: Cambridge University Press.

Trevarthen, C., and P. Hubley. 1978. Secondary intersubjectivity: Confidence, confiding, and acts of meaning in the first year. In *Action, Gesture, and Symbol: The Emergence of Language,* ed. A. Lock, 183–266. London: Academic Press.

Triandis, H. C. 1995. *Individualism and Collectivism.* Boulder, CO: Westview.

Trimble, J. E., S. M. Manson, N. G. Dinges, and B. Medicine. 1984. American Indian concepts of mental health: Reflections and directions. In *Mental Health Services: The Cross-Cultural Context,* ed. P. B. Pedersen, N. Sartorius, and A. J. Marsella, 199–220. Beverley Hills, CA: Sage.

Troje, N. F. 2013. What is biological motion? Definition, stimuli, and paradigms. In *Social Perception: Detection and Interpretation of Animacy, Agency, and Intention,* ed. M. B. Rutherford and V. A. Kuhlmeier, 13–36. Cambridge, MA: MIT Press.

Trommsdorff, G., and W. Friedlmeier. 2004. Zum verhaltnis zwischen kultur und individuum aus der perspektive der kulturvergleihenden psychologie. In *Positionen der kulturanthropologie,* ed. A. Assmann, U. Gaier, and G. Trommsdorf, 358–386. Frankfurt: Spektrum.

Trommsdorff, G., and H. J. Kornadt. 2003. Parent-child relations in cross-cultural perspective. In *Handbook of Dynamics in Parent-Child Relations,* ed. L. Kuczynski, 271–306. London: Sage.

Uller, T. 2008. Developmental plasticity and the evolution of parental effects. *Trends in Ecology and Evolution* 238:432–437.

Umilta, C., F. Simion, and E. Valenza. 1996. Newborns' preferences for faces. *European Psychologist* 1 (3): 200–205.

Valsiner, J. 1984. Construction of the zone of proximal development in adult-child joint action: The socialization of meals. In *Children's Learning in the "Zone of Proximal Development,"* ed. B. Rogoff and J. V. Wertsch, 65–76. San Francisco: Jossey-Bass.

Vandekerckhove, M., C. von Scheve, S. Ismer, S. Jung, and S. Kronast. 2008. Regulating emotions: Culture, social necessity, and biological inheritance. In *Regulating Emotions: Culture, Social Necessity, and Biological Inheritance*, ed. M. Vandekerckhove, C. von Scheve, S. Ismer, S. Jung, and S. Kronast, 1–12. Oxford: Blackwell.

Vihman, M. M. 1996. *Phonological Development: The Origins of Language in the Child.* Oxford: Blackwell.

von Hofsten, C., and K. Rosander. 2013. The establishment of gaze control in early infancy. In *The Development of Sensory, Motor, and Cognitive Capacities in Early Infancy: From Sensation to Cognition*, ed. F. Simion and G. Butterworth, 49–66. Hove, England: Psychology Press.

Vygotsky, L. S. 1978. *Mind in Society: The Development of Higher Psychological Processes.* Cambridge, MA: Harvard University Press.

Walden, T. A., and A. Baxter. 1989. The effect of context and age on social referencing. *Child Development* 60 (6): 1511–1518.

Warneken, F., F. Chen, and M. Tomasello. 2006. Cooperative activities in young children and chimpanzees. *Child Development* 77 (3): 640–663.

Wertsch, J. V. 1998. *Mind as Action.* New York: Oxford University Press.

Wertsch, J. V. 2002. *Voices of Collective Remembering.* Cambridge: Cambridge University Press.

Wheeler, M. 2005. *Reconstructing the Cognitive World.* Cambridge, MA: MIT Press.

Wilson, A., and C. F. Craver. 2007. Realization: Metaphysical and scientific perspectives. In *Handbook of the Philosophy of Psychology and Cognitive Science*, ed. P. Thagard, 81–104. Burlington: Elsevier.

Wilson, D., and D. Sperber. 2002. Truthfulness and relevance. *Mind* 111 (443): 583–632.

Wilson, D., and D. Sperber. 2004. Relevance theory. In *The Handbook of Pragmatics*, ed. L. Horn and G. Ward, 607–632. Oxford: Blackwell.

Wilson, R. A. 2001. Two views of realization. *Philosophical Studies* 104 (1): 1–31.

Wilson, R. A. 2004. *Boundaries of the Mind: The Individual in the Fragile Sciences—Cognition.* Cambridge: Cambridge University Press.

Wilson, R., and A. Clark. 2009. How to situate cognition: Letting nature take its course. In *The Cambridge Handbook of Situated Cognition*, ed. P. Robbins and M. Aydede, 55–77. Cambridge: Cambridge University Press.

Wilson, R. A., and L. Foglia. 2011. Embodied cognition. In *Stanford Encyclopedia of Philosophy*, ed. E. N. Zolta. http://plato.stanford.edu/archives/fall2011/entries/embodied-cognition/.

Winsler, A., R. M. Diaz, and I. Montero. 1997. The role of private speech in the transition from collaborative to independent task performance in young children. *Early Childhood Research Quarterly* 12:59–70.

Wood, D., J. Bruner, and G. Ross. 1976. The role of tutoring in problem solving. *Journal of Child Psychology and Psychiatry, and Allied Disciplines* 17:89–100.

Wray, A. 2000. Holistic utterances in protolanguage: The link from primates to humans. In *The Evolutionary Emergence of Language*, ed. C. Knight and M. Studdard, 285–302. Cambridge: Cambridge University Press.

Wray, A. 2002. *Formulaic Language and the Lexicon*. Oxford: Oxford University Press.

Wyman, E., and M. Tomasello. 2007. The ontogenetic origins of human cooperation. In *The Oxford Handbook of Evolutionary Psychology*, ed. R. Dunbar and L. Barrett, 227–236. Oxford: Oxford University Press.

Yap, P. M. 1965. Koro: A culture-bound depersonalisation syndrome. *British Journal of Psychiatry* 3:43–50.

Zac, P., W. Matzner, and R. Kurzban. 2008. The neurobiology of trust. *Scientific American Magazine* 298 (6): 88–95.

Zajonc, R. B. 1980. Feeling and thinking: Preferences need no inferences. *American Psychologist* 35 (2): 151–175.

Zajonc, R. B., S. T. Murphy, and M. Inglehart. 1989. Feeling and facial efference: Implications of the vascular theory of emotion. *Psychological Review* 96 (3): 395–416.

Zinck, A., and A. Newen. 2008. Classifying emotion: A developmental account. *Synthese* 161 (1): 1–25.

Index